Springer Praxis

Popular Scie

GW01395983

The Springer Praxis Popular Science series contains fascinating stories from around the world and across many different disciplines. The titles in this series are written with the educated lay reader in mind, approaching nitty-gritty science in an engaging, yet digestible way. Authored by active scholars, researchers, and industry professionals, the books herein offer far-ranging and unique perspectives, exploring realms as distant as Antarctica or as abstract as consciousness itself, as modern as the Information Age or as old our planet Earth. The books are illustrative in their approach and feature essential mathematics only where necessary. They are a perfect read for those with a curious mind who wish to expand their understanding of the vast world of science.

Roberto Manzocco

Genius

Theory, History and Technique

Published in association with

Praxis Publishing
Chichester, UK

Springer

PRAXIS

Roberto Manzocco
Department of Anthropology
John Jay College of Criminal Justice
New York City, NY, USA

Springer Praxis Books
ISSN 2626-6113   ISSN 2626-6121   (electronic)
Popular Science
ISBN 978-3-031-27091-8   ISBN 978-3-031-27092-5   (eBook)
https://doi.org/10.1007/978-3-031-27092-5

© The Editor(s) (if applicable) and The Author(s), under exclusive license to Springer Nature Switzerland AG 2023
This work is subject to copyright. All rights are solely and exclusively licensed by the Publisher, whether the whole or part of the material is concerned, specifically the rights of translation, reprinting, reuse of illustrations, recitation, broadcasting, reproduction on microfilms or in any other physical way, and transmission or information storage and retrieval, electronic adaptation, computer software, or by similar or dissimilar methodology now known or hereafter developed.
The use of general descriptive names, registered names, trademarks, service marks, etc. in this publication does not imply, even in the absence of a specific statement, that such names are exempt from the relevant protective laws and regulations and therefore free for general use.
The publisher, the authors, and the editors are safe to assume that the advice and information in this book are believed to be true and accurate at the date of publication. Neither the publisher nor the authors or the editors give a warranty, expressed or implied, with respect to the material contained herein or for any errors or omissions that may have been made. The publisher remains neutral with regard to jurisdictional claims in published maps and institutional affiliations.

This Springer imprint is published by the registered company Springer Nature Switzerland AG
The registered company address is: Gewerbestrasse 11, 6330 Cham, Switzerland

Genius is 1% inspiration and 99% perspiration
—Thomas A. Edison

Chance favours the prepared mind
—Louis Pasteur

This book is dedicated to Pietro Greco

Introduction

It Doesn't Take a Genius

Who knows how Einstein must have felt—a man literally capable of lifting, even if only slightly, the veil that separates us from full knowledge of reality and, in so doing, shattering the world of physics—during his daily interactions with other people, during his daily tasks, and so on? More generally, what does it mean to penetrate an area of reality and knowledge to such an extent that you become capable of going beyond it? And how does it change you to entertain, for days, years, and decades, a relationship with a phenomenon, a concept, or a work of art that absorbs you completely, to the point of identifying it with your very existence? What does it mean, in essence, to be a genius?

Fortunately, writing a book on genius does not require one to be a genius, because, among other things, we can apply the famous advice of Newton, who famously said in a letter to his rival Robert Hooke in 1676: "If I have seen a little further it is by standing on the shoulders of giants." When we began the preparatory work for the book you are about to read, we certainly did not expect to discover the exaggerated amount of literature produced over the years—indeed, centuries—on this subject, and especially the high quality of many of the contributions we found ourselves reading. In short, we have at our disposal many sets of shoulders of giants who, before us, have dealt with these things on a technical level, deducing and identifying all kinds of conceptual subtleties in the study of genius. Without the work of these scholars, this book—which serves as a popular introduction to these issues—would not have been possible.

Among the outstanding authors who have lent their shoulders to our pen, we would like to mention, first and foremost, Dean Keith Simonton, a scholar who has not only developed a very original perspective of genius, but has also systematized and analyzed in detail all of the scholarly literature relating to this subject; Nacy Andreasen, who has explored the neuroscience of genius; and K. Anders Ericsson, who sought to identify the winding path that greats in every field must pursue to achieve excellence.

In Chap. 1, we will deal with the history of the concept of genius, i.e., how this idea was born and how it has been transformed over the centuries. In Chap. 2, we will examine some psychological constructs akin to genius, i.e., intelligence, creativity, charisma, and whatnot—the perspective we will try to follow on the subject of genius will, in fact, always be related to neuroscience and psychology, rather than philosophy. In Chap. 3, we will explore the Darwinian theory on the origin of genius elaborated by Simonton, Campbell, and other authors; in Chap. 4, we will attempt to examine what happens in the brains of genius people. In Chap. 5, we will review all of the open questions related to genius, focusing, in particular, on the combination of genius-madness and nature-nurture. In Chap. 6, we will explore genius in the specific domain of the natural sciences, while in Chap. 7, we will explore Ericsson's work and, more generally, the path one has to take in order to rise to the level of genius. Chapter 8, which is shorter, will be devoted to machines and the speculation concerning the possibility of creating synthetic genius minds. As you may have guessed, this is an introductory text, born with the hope of enticing you to dive into the concluding bibliography and delve deeper into these fascinating topics, in the knowledge that it is a path without end.

Contents

1

Clearing Up the Sky. Sages, Heroes, Saints, and Geniuses

1.1   The Source. Rome and Greece

Imhotep—'he who comes in peace'—was an official of Pharaoh Djoser, and the architect who built the Step Pyramid of Saqquara. He was also high priest of the Sun-God Ra at Heliopolis. Given that he lived towards the end of the twenty-seventh century BC, not much is known about him from a historical point of view, except that he was later deified over a period of time. Imhotep is probably the first genius that we can historically identify. Not that the term 'genius' was applied at the time. For that to happen, we have to wait for the Roman era, but for the word to take on its modern meaning, we have to wait for the era from which this adjective derives—'modern' indeed.

The Modern Age, for its part, has done away with a whole series of inter-mediaries that traditional culture offered us common mortals in order to bring us closer to the divine. Demigods, demons, angels, spectres, as well as saints, the blessed, martyrs and magicians. Heaven therefore remained empty, but not for long. What took the place of these characters, real and fictional, was a new type of figure, a set of exceptional men and women who coagu-lated around themselves a kind of informal religion, which we could define as 'the cult of genius'. Supporting this interesting thesis is a historian from Dart-mouth College (Hanover, New Hampshire), Darrin M. McMahon, who has

© The Author(s), under exclusive license to Springer Nature
Switzerland AG 2023
R. Manzocco, *Genius*, Springer Praxis Books,
https://doi.org/10.1007/978-3-031-27092-5_1

devoted an excellent book to the subject in question that we highly recommend, *Divine Fury*.[1] In the course of this chapter, we will rely heavily—albeit with a few distinctions—on McMahon's work.

The contemporary use of the label 'genius' dates back only a few centuries, but the fascination with human eminence is as old as humanity itself. In this first chapter, then, let us take a look at how this idea of genius developed and how we arrived at the present definition.

As far as classical literature is concerned, the term 'genius'[2] appears officially for the first time in a work by Plautus, a Roman playwright who lived between the third and second centuries before Christ; in the work *Captivii*, the author narrates, among other things, the story of a very miserly character who officiates at a sacrifice of little value to his own personal genius. Yes, because, in the ancient Roman religion, this is what genius is: an individual spirit present in every person—a bit like the guardian angel in the Christian tradition, who would follow us from birth to death-, but also in things and, especially, places (the so-called 'genius loci'). Originally, the 'genius,' somewhat like the 'mana' dear to anthropologists, constitutes a kind of elemental force that penetrates the world, a sacred power analogous to what the Romans called 'numen'. We do not know, however, when this primal force was transformed into a series of individual spirits acting as protectors. Be that as it may, the genius, or rather the geniuses, became, at some point, personal protectors associated with human beings from birth and responsible for the temperament of the latter, so much so that, in statuary representations, the geniuses are depicted with the features of their protectors. Another word associated with this one is 'ingenium,' which can be translated as 'natural disposition' or 'talent.' And the two concepts manifest a tendency to merge.

After all, the figure of the genius has something profoundly religious about it, even in modern times; just think of Einstein's well-known desire to know God's thoughts. Einstein, the secular saint capable of understanding transcendental truths. It is precisely for this reason that we are fascinated by genius in its contemporary sense: for the fact that it can grant some of us, and therefore, by derivation, all of us, direct access to the more obscure mechanisms that preside over the workings of the world. In other words, the genius can peer through the curtain that separates superficial reality from what lies behind it, revealing a world far more complex and strange than we expected.

One thinks, for example, of the work of the American critic Harold Bloom, who, in *Genius. A Mosaic of* 100 *Exemplary Creative Minds,* describes precisely a hundred geniuses as cabalistic representations of God.

---

[1] McMahon, Darrin, *Divine Fury: A History of Genius*, Basic Books, New York 2013.
[2] *Genius*, from the Latin verb *gigno, gignere*, meaning 'to generate'. Of obscure origin.

Even earlier, in 1931, the German psychiatrist Wilhelm Lange-Eichbaum, in *Das Genie-Problem*, speaks clearly of the tendency of the time to deify the figure of the genius. The author, however, is not particularly eager to submit to this cult, for that is what it is; and declaring this in the same year is the American historian and philosopher Will Durant, in his *Adventures in Genius*, according to which the religion of genius is the "final religion." In short: all holders of genius, of whatever kind—scientific, artistic, military or whatever else—have taken the place, traditionally belonging to the saints, of mediators between the human and the divine, a transition that took place around the eighteenth century, in the European West.

Despite the growing faith in the equality of all human beings from the eighteenth century onwards, despite the French and American Revolutions, the idea of an inequality based on intellect and creation was manifested in this era. Thomas Jefferson himself spoke of a 'natural aristocracy' that included people endowed with intellect, creative ability and talent. Precisely an aristocracy of genius, a concept that would persist to this day, and that ends up being confused with a different concept, that of celebrity, which will lead us to speak of genius in fashion, business, sport, and so on.

Admittedly, the concept of genius is a product of Western modernity, but this has not prevented other cultures from celebrating the mental abilities of, for instance, Brahmins, Buddhist monks, rabbis, and so on. Genius is a label that we tend today, therefore, with a certain degree of legitimacy, to apply retroactively to people from other eras and other cultures. With the risk of committing attribution errors, however: be careful not to confuse genius, whose main attribute is originality, with the figure of the sage, dear to many cultures—the Chinese, for example.

The cult that modern Europe reserves for the almost blasphemous originality of genius does not mean, of course, that there was no creation, innovation or imagination before now; however, previous eras, and, in particular, the Greek tradition of *hybris* and the Judeo-Christian tradition of a single creator God, caused originality to be somewhat sidelined, promoting the idea that there is nothing new under the Sun and that everything is an imitation of something else. Since to disagree with this view entails an accusation of hubris, arrogance and blasphemy, one can understand how courageous the new vision of human creativity that has emerged in modern Europe is. The modern cult of genius is, in some respects, a direct attack on God.

Since, then, willingly or unwillingly, we are forced to trace this history of genius before the current use of the term in question, from what or whom do we begin? From the wisest of mortals, McMahon suggests, from that Athenian philosopher who proclaims he knows nothing: Socrates, indeed. Right now, though, we are not so much interested in his philosophy as in another aspect of his life. That is, his daimonion, that entity of divine origin that, according to him, holds him back from doing something he set out to do. The *daimonion,* diminutive of *daimon,* from which we derive our 'demon'.

It is not such a strange idea. After all, in that era, the idea of minor spirits acting as guardians was quite widespread among ordinary people. These demons represented a kind of intermediary between the world of humans and that of the gods proper, and werew also used to explain insights and abilities beyond the ordinary. In fact, these demons also acted as possessors of people with poetic and artistic abilities, which would therefore not be considered innate, but to have come from "elsewhere." And the poet would therefore simply be a passive receptor of the messages of the Muses, who possess him or her—an idea supported, for example, by Plato himself, in dialogues such as *Ion* and *Phedrus,* which were very influential on the later interpretation of the nature of genius. In short, the Platonic idea would only come to settle into later thought, coagulating into the Latin idea of 'furor poeticus.' Socrates' disciple thus rails against the earlier Greek idea of 'technê,' of practical skill in need of cultivation, a trait that contemporaries also attributed, at least in part, to poetry. For philosophy, history is different: it can be learned, that is, one can be trained to be a philosopher. The concept of poetic fury would be opposed by Plato's most famous disciple, Aristotle, who saw the mind as the source of his own imaginative abilities.

If, a hundred years before Plato, the poet Pindar had compared innate capacities with acquired capacities and argued in favour of the former—identifying nature and inspiration, thus not seeing them as opposites-, Aristotle, in his *Poetics,* rather tries precisely to distinguish natural capacities and inspiration. In particular, the Stagirite contrasts poetry as a gift of nature with that which is instead produced by a "strain of madness" that drags the poet "out of his proper self."

1.2   The Christian Era and Modernity

In any case, in the Christian era, these intermediary figures are replaced by characters of a different type: first of all, the martyrs, and later the saints, who thus assume the role that belonged to the Roman genii, as witnessed by art and iconography, as well as popular devotion.

But if there is one figure that, in the Christian era—and particularly from the Renaissance onwards—will influence the later figure of the genius, it is that of the magician. Take, for instance, the Germanic myth of Faust, who sells his soul in exchange for knowledge—inspired by a real-life magician from the fifteenth century, the legend later inspired famous works by authors such as Johann Wolfgang von Goethe and Christopher Marlowe. Myths about magicians and their sometimes fatal fate represent a warning about the dangers inherent in the quest for ultimate knowledge.

To remain in the Renaissance, let us also mention Marsilio Ficino, for whom the only intermediary between humanity and God is the soul. Humanity is, in essence, the median point of the Universe, and, given this role, they can choose whether to live vegetatively, like a plant, sensually, like an animal, or intellectually, like an angel. Not only that: poets and intellectuals would be inspired by God Himself, and could use this inspiration to ascend directly to His level.

As far as the sixteenth century is concerned, we must at least mention Juan Huarte, who dedicated a treatise to the subject of genius, *Examen de Ingenios* (1575). Taking up the poetic fury described by Plato, the scholar denies its divine origin and tries to develop a physiological explanation for it, using the conceptual tools available at the time, in particular, by referring to the classical theory of humours. Not that Huarte considers genius to be something original: only God can be that, and all human work is but imitation. This fact, however, did not prevent Huarte's contemporaries, or rather, his immediate predecessors, from sanctifying, for example, an eminent artist such as Michelangelo, upon whose death rumours began to circulate concerning the incorruptibility of the body and its alleged thaumaturgic virtues.

This is not the first time that the label 'divine' had been used for an artist: just think of Dante, Petrarch, Brunelleschi. The idea that these figures should be seen as geniuses, however, dates back to Romanticism, which reinterpreted them as 'geniuses'. It must be said, however, that the Renaissance began the process of cleansing celestial space that would free Western civilisation of all of the intermediaries already named, leaving room for geniuses. The humanist Julius Caesar Scaliger, for example, speaks of the poet as an 'alter Deus,' another God, while in the sixteenth century, Torquato Tasso argues that there are two creators, God and the poet.

In the 1650s, the English philosopher Thomas Hobbes criticised the idea that artistic and intellectual ability derives from God, the Muses or some other form of supernatural possession, as if the creator subject was actually a kind of bagpipe—a Hobbesian metaphor—played by external forces. In short, for Hobbes, the source of his writings are his personal meditations, not 'enthusiasm'—being 'full of God,' a term derived precisely from the Greek 'en' and 'theos,' coined there in that period and since having become very popular, in a pejorative sense, however, often used to indicate religious fanaticism.

The seventeenth century also means John Locke and the empiricist philosophy derived from him, and thus the doubt cast by it on the concept of divine inspiration. Empiricism and criticism of enthusiasm thus give rise to a systematic attack on the mystery of genius. Among the most prominent critics—and, by now, we are already in the eighteenth century—are two intellectuals influenced by Locke's empiricist psychology and epistemology, the Anglican cleric William Sharpe—*Dissertation upon Genius*, 1755—and the French philosopher Claude Adrien Helvétius—in several sections of *On the Mind*, 1758. For Sharpe, the mind being a 'blank slate', everything is the product of education, which is imprinted on it, and this includes genius, which is not the product of nature, but of the vicissitudes of life. According to Helvétius, all human beings are endowed with sufficient memory to enable them to acquire the highest human abilities. It follows that the origin of excellence lies not in birth, but in what happens afterwards, education, the vicissitudes of life, training, luck, and so on. What is important here is that both Sharpe and Helvétius try to eliminate the mystery of genius by showing that genius people have nothing that is qualitatively different from what other people have. French philosophers such as Condillac, Turgot, d'Alembert, and Condorcet also promoted this conception, not coincidentally at a time when a profound faith in human equality was also affirmed.

These authors and this era were not exactly comfortable with the definition of the concept of genius, which was elitist by nature. This difficulty was expressed in 1711 by the English scholar Joseph Addison in an article published in the *London Spectator*. In it, Addison distinguished between two types of genius: natural geniuses, who, by nature, break the rules and produce great works—Homer, Shakespeare, and so on—and imitative geniuses, who become such through study and effort and submit to the rules of the Art they serve—e.g., Aristotle, Virgil, Milton.

And the poet Edward Young, in his work *Conjectures on Original Composition* of 1759, distinguishes between what is acquired through learning and what is produced by supernatural inspiration, an anticipation of the distinction made later between natural talent, developed through learning, and genius, granted by nature to a privileged few. This brings us all the way to the philosopher Arthur Schopenhauer, with his famous distinction—in *The World as Will and Representation*—between talent, as the ability to hit a target that others cannot reach, and genius, as the ability to reach a target that others cannot even see. These authors therefore work in the opposite direction to those who seek to 'naturalise' genius and reduce it to mere training. In short, the eighteenth century witnessed a gradual shift from a naturalised view of genius to the idea that human ingenuity is a unique phenomenon that cannot be reduced to factors such as acquired mastery.

And this century also saw the rehabilitation of the term 'enthusiasm,' which took on a decidedly positive connotation. The person of genius feels the presence of the flame of enthusiasm that pervades their soul, observes the Swiss philosopher and mathematician Johann Georg Sulzer, for example, while the German critic Carl Friedrich Flögel argues that genius is a fire that burns because of the enthusiasm that permeates it. In short, the language used by many of these authors is full of metaphors referring to possession in order to convey the idea of a genius that is unique and somehow not reducible to rationality and materialism. Also resorting to such transcendent language is Diderot, who, despite his atheism, speaks in poetic terms of enthusiasm—it is true that the French thinker also tries to give a description in physiological terms of poetic fury, which, however, in turn, retains its simil-religious peculiarity, namely, the ability to transcend human limits and reveal aspects of the world hidden from the common mortal eye. Even Kant, albeit with all of the limitations of the case, recognises—in the *Critique of Judgement*—that genius must be precisely original, and therefore contrary to the spirit of imitation.

Thus, the eighteenth century, or at least some of the souls of that century, went against the thinking of previous centuries, obsessed with the idea of imitation, and dared to suggest that genius could somehow share creative capacity with God himself. And so, if, as we have seen in the Renaissance, Julius Caesar Scaliger and a few others posited the poet as a kind of other divinity, in the early eighteenth century, the English politician and philosopher Anthony Ashley Cooper, Third Earl of Shaftesbury, argued that the true poet is a second Maker. All statements that, at the time, could easily have produced accusations of blasphemy and the like.

Given this process of exaltation of genius and occupation of an empty sky, one should not be surprised, therefore, at the reception received by a child prodigy like Mozart. Helvétius describes him as one of the most singular beings in existence, and a member of the British Royal Society, Daines Barrington, studied the boy—who was then eight years old—and was able to ascertain Mozart's improvisational abilities, which proved his genius beyond reasonable doubt.

Famous are the philosopher Immanuel Kant's reflections on the subject of genius, produced, according to him, by the "hand of nature." A mysterious quality, however, that he attributes only to artists, not scientists. And, as an example, Kant brings up the case of Isaac Newton, whose work can be studied, taught and, above all, understood step by step. So, it is not generic excellence that is mysterious, but the creative process of the arts. In upholding this position, Kant goes against the spirit of his age, which conceived of genius in rather broad terms. And indeed, despite Kant's assertions, Newton was seen by many as a genius at the time.

Also dating from this period is *The Nephew of Rameau*, a work by Denis Diderot composed in the 1760s—and translated into German in 1805 by Goethe. It is a dialogue between Jean-Philippe Rameau and a philosopher who more or less represents the author himself, emphasising the 'dark side' of genius. In particular, the protagonist declares that evil always comes into the world because of some man of genius. Geniuses are hateful people, bad parents, bad brothers and bad friends. Although they can change the world, Rameau continues, he would prefer the world to be without them. Geniuses are monsters, in essence. Their originality and essential loneliness set them against society itself, which admires them but also fears them; they are a destabilising force, which can lead to madness and crime.

Meanwhile, on the other side of the Atlantic, having risen to the level of a genius thanks to his experiments on electricity, Benjamin Franklin—far from embodying the evil genius envisaged by Rameau—became the object of a veritable cult of his image, printed almost everywhere, on medallions, rings, busts, prints. His fame preceded him even to France, where Franklin went as ambassador, and where he was treated with great reverence.

In 1795, the Prussian scholar Alexander von Humboldt gave us a short story known as *The Vital Force, or The Rhodian Genius*, set in Syracuse in the fourth century BC. In the work, questions are asked about the nature of genius,BNS which is explained as a kind of vital energy, of which the man of genius would be, in a sense, 'overloaded', an idea that relates back to the all-German concept of *Lebenskraft*—a term we can translate as 'vitality'. It was a very popular concept at the time, that of a vital energy responsible for, among other things, interpersonal influence, genius and charisma. For example, the American politician and writer Fisher Ames argued, at one point, that genius is to the intellectual world as electricity is to nature, while writer Mary Shelley—you remember her, right?—tells us the story of an evil genius who uses electricity itself to capture the principle of life, and thus becomes a "modern Prometheus." Ironic, isn't it? Authors who use the scientific knowledge of the time to characterise genius, however, end up returning to a mystical vision of genius, understood precisely as an almost supernatural force that is difficult to understand, like 'mana' or 'numen.'

The era in question also saw the rise of a historical figure whose importance is difficult to diminish, whom Goethe would describe as a genius and whose bust Byron kept in his room, infuriating his classmates at Harrow School: Napoleon Bonaparte. IfBNS, generally, in the eighteenth century, the title of genius was attributed mainly to scientists and artists, it is true that sometimes even people like Richelieu, Frederick the Great or George Washington were considered such. In the case of Napoleon, his genius was considered by his contemporaries then as the very reason for his rise to power. Not to mention the fact that he was, following his death, even given a sort of cult of relics and memorabilia.

In the midst of the Romantic era, the question arises once again as to what this mysterious genius that some human beings possess consists of. If the philosopher Johann Gottlieb Fichte argues with Kant that genius is a force that cannot be described, for Friedrich Schelling, the concept of genius is similar to that of Destiny, which realises unpredictable purposes through us. But, having gotten angels, spirits and Muses out of the way, what is left for thinkers of the Romantic era to explain genius? While Percy Shelley refers to a poet-prophet who can intuitively connect with a universal mind, August Wilhelm Schlegel—Friedrich's brother—in his *Lectures on Dramatic Art and Literature* (1809–1811), argues that, in essence, the activity of genius is unconscious. In short, positions vary, but the central idea always remains more or less the same: genius is such because, consciously or unconsciously, it manages to 'participate' in that which is eternal and infinite. Shelley sees it neo-Platonistically as a 'universal mind', others refer to the Absolute, the Will, or the Hegelian 'Spirit,' while there are also those who refer to Spinoza's pantheism and those who continue to speak of the more traditional God.

But, in the Romantic era, there is another theme, which also constitutes the not-so-secret fear of every person aspiring to produce art and knowledge: that of the genius misunderstood and unrecognised in life. The concept of forgotten, unrecognised genius was a typical theme of the Romantic movement, as well as a widespread fear among these same authors. A fear that, far from disappearing, persisted in the following decades, with the exemplary case being that of Vincent Van Gogh. A myth that contrasts with the other romantic conception of genius as epiphany, as a force that imposes itself on history.

1.3   The Scientific Assault on Genius

Alongside the Romantic current, other currents of thought of a completely different nature were trying to have their say on the subject of genius, framing it in a theoretical context that may make one smile today, but which, at the time, seemed to make complete sense. We cite, in particular, physiognomy—the study of human character from the face—and phrenology—according to which intellectual faculties and character traits could be deduced from the shape of the skull. Could these two disciplines—nowadays fallen into complete disrepute—refrain from having their say on genius? Certainly not.

A Swiss Protestant pastor, Johann Caspar Lavater, was the creator of physiognomy, a discipline that exerted a broad influence on the study of genius, beginning with the work of the founder himself, who, in his extensive work *Physiognomische fragmente*—published between 1775 and 1778—tells us that a power such as that of genius can only manifest itself in the flesh. For Lavater, genius is, first and foremost, seen in the eyes, but also in the physical form. The 'intensive' geniuses—that is, those focused on a single subject—would have strong bones, slower and heavier movements, and pronounced foreheads; the 'extensive' ones, on the other hand, would have elongated faces and a more delicate physical structure.

This approach was later replaced with a method considered more 'objective' at the time, developed by the German Franz Joseph Gall, the 'cranioscopy', later renamed 'phrenology'—from the Greek 'phren,' 'mind,' and 'logos'—by his collaborator and pupil Johann Caspar Spurzheim. Between the eighteenth and nineteenth centuries, attention therefore shifted from the face to the skull, understood as the external expression of the brain, the true source of genius. An idea that, given its materialistic nature, could only anger the Catholic Church, resulting in Gall's expulsion from Vienna.

According to the father of phrenology, it would be possible to identify, from the skull, attributes such as a propensity towards theft or murder, musical sense and technical skills, moral sense and parental love, and much more. Thus also exceptional abilities, and thus genius. While Gall makes room for different types of genius, connected to specific disciplines, his successors tried to identify a kind of generic genius, a 'thing' that is held in common by great scientists and artists, leaders and politicians. A passionate collector of skulls, Gall developed an extensive collection, which, at the time of his death (1828), consisted of the heads of 103 notables, 69 criminals, 67 mentally ill persons, and more. Not only that, but, true to his own doctrines, he also left his own head to posterity. The spread of phrenology brought with it a wave of thefts and the trafficking of 'excellent' skulls. At the death of the famous composer Joseph Haydn, an admirer of Gall's work bribed one of the funeral attendants and obtained the musical genius' skull, which then passed from hand to hand until it was returned to the deceased owner after the Second World War. Other geniuses found themselves losing their heads as well: for example, the Spanish painter Francisco Goya, the Swedish mystic Emanuel Swedenborg, and even Beethoven, whose skull would be divided into parts. A trade that can only remind us of the most famous medieval trade in saints' relics.

And the passion of scientific research for the heads, skulls and brains of eminent figures showed no sign of disappearing, even after the decline of phrenology. And so, for example, in 1855, the German anatomist Rudolf Wagner managed to get his hands on the brain of the great mathematician Karl Friedrich Gauss. Illustrious 19th-century scholar of the human brain, Frenchman Paul Broca, created the then-largest collection of brains and skulls in the world, more than seven thousand specimens, pointing out, above all, their size and weight, qualities considered to stand in relation to intellectual capacity. Yes, because, in this phase of neurology, size matters: in particular, let us mention the Russian writer Ivan Turgenev, whose brain weighed more than two kilograms, a record, but not enough to definitively connect genius and brain size. Take, for example, Gall's paltry 1198 g, and Broca's own 1424 g. For his part, he founded a Society of Mutual Autopsy, under which he and other eminent French citizens undertook to donate their heads to science for the study of the enigmas of the genius brain.

Physiognomy, phrenology and neurology, then. But also psychiatry. For the French scholar Louis-Francisque Lélut, the key to genius is, in fact, pathology. In 1836, the psychiatrist published a book, *On the Demon of Socrates*, in which he argued that one of the greatest philosophers in history suffered from hallucinations. Lélut's book is one of the first instances of 'pathography', i.e., a medical biography of the pathologies that afflict geniuses—for the record, this term was later coined by the German neurologist Paul Julius Möbius, grandson of August Ferdinand Möbius, father of the famous and eponymous 'strip.' In 1846, Lélut published another book in which he claimed that Blaise Pascal also suffered from this hallucinatory condition.

Even for the psychiatrist Jacques-Joseph Moreau—among other things, a founding member, together with Théophile Gautier, of the Club des Hashischins—the source of genius is mental illness. The club in question included the likes of Baudelaire, Victor Hugo, and Gérard de Nerval. The club's 'smokes' impressed Moreau, for, with them, apparently healthy people could experience and express altered states of consciousness similar to those of the mentally ill. The link between genius and mental illness is, according to the psychiatrist, so strong that he coined a special term, *pathogénie*—which stands for pathological genius. In essence, for Moreau, geniuses are born, not made, and genius is a semi-pathological state of the brain, no more and no less. So, between the genius who questions the origin of the ideas that come into their head and the mentally ill person who believes they are talking to superior entities, the difference is only one of degree. The line of enquiry opened up by Moreau would continue in the work of the German psychologist Ernst Kretschmer, who, in 1929—in his work *Psychology of Men of Genius*—argues that, in order to transform talent into true genius, it is necessary to add "a psychopathic element."

Before we move into the twentieth century, however, we must still talk about Cesare Lombroso. Known for his work on criminal anthropology, the scholar actually showed a strong interest in the subject of genius right from the start, so much so that he dedicated one of his first books to it, *Genius and Madness* (1864), later revised and republished in 1889 as *The Man of Genius*. He even created a neologism, 'devolutionism,' with which he indicated his central idea, namely, that human beings can, in certain cases, manifest primitive tendencies and forms of involution. In short, there are 'low' involutionary degenerative tendencies—mental illnesses and anti-social behaviour—and 'high' forms, which are present in the geniuses, such as emotional instability, selfishness, moral absence and impulsiveness. Obviously, the 'devolution' would manifest itself as much in the mind as in the body, so much so that the geniuses would frequently be small in stature, emaciated, pale, often left-handed, sickly when young, and bearing little resemblance to their parents. To support his ideas, Lombroso also resorted to craniometric analyses. Even traumas do not escape Lombroso's notice; for example, the scholar cites the story that the Italian philosopher Giambattista Vico owed his intellect to a fall that he suffered as a child.

And here, finally, comes Francis Galton into the picture. An English polymath, Darwin's cousin and the father of eugenics, Galton dealt with the subject of genius in purely scientific terms for a long time. In 1865, he published an article entitled *Hereditary Talent and Character*, the starting point for a larger work (of 1869) that became very famous, *Hereditary Genius: An Inquiry Into Its Laws and Consequences*. This work is considered to be the first scientific study ever undertaken specifically on genius. From the very title, the work suggests Galton's approach to the subject, i.e., the idea that genius is a hereditary factor, a thesis he sets out to prove through genealogical analysis of several intellectually eminent families in the United Kingdom.

To carry out his study, Galton turned to the approach developed earlier by the Belgian mathematician Alphonse Quételet, who had made use of quantitative statistical methods to demonstrate, among other things, that the pectoral dimensions of Scottish soldiers followed a law of deviation from a mean that, translated into graphical terms, resembled a Gaussian curve—what we now call the Bell Curve. The idea promoted by Quételet and Galton is that nature tends to coalesce around a mean, with a symmetrical distribution on both sides. And, if this is true for physical characteristics such as chest size, height and other physical characteristics, then it must also be true for intellectual abilities. For Galton, true geniuses are rare: one in a million, to be precise, the scholar tells us at one point. But what criteria does he use to determine whether or not someone belongs to this select group? Only one, a somewhat questionable one, in fact: reputation. Elevated reputation would, according to him, mean elevated ability. In particular, Galton writes of a "triple concrete event"—i.e., natural ability, dedication and the ability to work hard—as the recipe needed to produce genius. And, despite romantic complaints about misunderstood geniuses, for Galton, if someone does not produce something that society classifies as 'eminent,' then this someone is not a genius. Full stop. So, between nature and nurture—a terminological juxtaposition he himself introduced—nature wins hands down.

Galton's work would greatly influence the culture of the time, inspiring the production of other later studies, such as Havelock Ellis's *A Study of British Genius* in 1904 and James McKeen Cattell's *American Men of Science* in 1906.

And this despite the poor results reported. In fact, from a certain point of view, the application of statistical methods only represents the recognition of something that is already known, not its explanation, just as the craniometric studies that were in vogue until recently did not constitute an explanation. Not to mention the fact that the study on eminent personalities concerns characters who are mostly already dead, while, as to the living, it does not have much to say. In particular, it does not allow us to predict who will become a genius, i.e., it does not allow us to identify potential geniuses. Galton therefore turns to the study of the living, collecting data on a variety of factors, such as handshake, head characteristics, reaction times, in the hope of identifying factors that can precisely predict future genius.

1.4 The Age of IQ

At this point, however, the baton passes to two other very important figures, who will be at the origin of studies on a subject that is still topical today, that of IQ. The first version of the relevant test was developed by the Frenchman Alfred Binet, with the aim of detecting not genius, but mental underdevelopment. This was at the request—in 1904—of the French government, which wanted to identify students with mental disabilities. This requirement gave rise to a classificatory terminology that would enter common language in the form of an insult: in fact, alongside 'normality', Binet and his colleague Théodore Simon would add the categories of 'idiocy', 'imbecility,' and more.

The purpose of the test is to compare the subject's mental age with his or her registry age by assigning him or her a series of tasks that would be considered simple if performed by a child of this or that age. And, in 1912, the German psychologist William Stern proposed dividing mental age by registry age and multiplying the result by one hundred, thus obtaining the IQ as we know it today.

And now, the second character in our story, after Binet, enters the scene: Lewis Terman. A professor of psychology at Stanford, in 1916, he published the Stanford Revision of the Binet-Simon, which was to become paradigmatic in the field of IQ measurement. For Terman, as for Galton, genius can only be based on heredity, and to this end, the American scholar set out to realise the dream of the English polymath, namely, the development of a method to identify geniuses before they manifest their excellence. After all, the future of human civilisation would be at stake, according to Terman. The first volume of his seminal work *Genetic Studies of Genius* came out in 1925, and summarised the results of extensive research that had begun in 1911. With it, Terman and his colleagues set out to identify as many small geniuses as possible in the school districts surrounding Stanford and San Francisco. Basically, the scholars administered their Stanford-Binet test to a large number of children, identifying about a thousand gifted children— later affectionately nicknamed 'Termites'-, all characterised by an IQ above 140 points, who would then be interviewed several times throughout their lives, at regular intervals. Needless to say, not only did few of the Termites attain a high level of intellectual eminence, but Terman's decades-long study discarded two children, Luis Alvarez and William Shockeley—owing to their IQs being below 140—each of whom would go on to win the Nobel Prize in Physics. The purpose of Terman's study was to identify genius, when what he actually achieved, at least in theory, was the identification of intelligence, a conceptual construct of a different kind—as we shall see. This term, like that of genius, is also ancient and laden with implicit meanings. One thinks of the use made of it by medieval theologians and philosophers specifically to indicate the Angelic Intelligences. In eighteenth-century taxonomy, the term 'intelligence' is extended to all living beings, which are precisely classified as more or less intelligent. To return to our story, it is no coincidence that, after 1945, Terman stopped using the term 'genius' to refer to the core of his work when he spoke about it in the press, while, in the fourth volume of *Genetic Studies of Genius,* published in 1947, the author admits the existence of a discrepancy between intellect and achievement. The real object of Terman's research then turns out to be 'general intelligence,' a concept we owe to another scholar, the British psychologist and statistician Charles Spearman;

the latter, in the early twentieth century, became convinced of the existence of a single factor, which he called 'g,' governing all of the different mental abilities measured by intelligence tests, from abstraction to verbal abilities, and so on. And so, just like many proponents of the study of genius, Spearman was also convinced of the existence of a single 'thing', the 'general intelligence,' in fact, prefixed, constant throughout life, and yet multiform. In this, he echoes Goethe, who considered genius—using modern terminology—to be something 'generic,' common to all eminent people, regardless of their speciality.

It should be noted that the second volume of *Genetic Studies of Genius* is not, however, devoted to the living, but to the dead: it contains the research of Terman's assistant, Catharine Morris Cox, who was, in fact, concerned with developing a method—debatable as it may be, and, indeed, subsequently the subject of much discussion—to measure the IQ of people who are no longer alive, starting with that of Galton, a genius himself, and the subject of a 1917 paper by Cox. With rather flattering results for him: around 200 points, a number based on analysis of available accounts of the activities Galton engaged in during his childhood, pastimes and whatnot.

Using the available historical accounts, and, in particular, those relating to a thousand eminent persons listed in the biographical dictionaries compiled by the American psychologist James McKeen Cattell, Cox first of all discards all characters whose childhood data is not available, thus eliminating those who lived before the sixteenth century. This leaves 301 characters, whose intelligence Cox assesses with her method. At the top, we find John Stuart Mill, with a handsome quotient of 190, immediately followed by Goethe, Leibniz and Hugo Grotius—all three with 185. Voltaire gets away with a 170—not bad, after all—while far lower scores go to such notables as Newton (130), Napoleon (135) and Beethoven (also 135).

1.5 The Cult of Genius: Nazifascism and Communism

In the meantime, another aspect of genius, or rather of its cult, flourishes in Europe; this concept is, in fact, regimented and put at the service of two movements that undoubtedly have mystical-religious aspects, namely, fascism and communism. This discourse on the cult of genius in modern Europe was first introduced by a Jewish historian, Edgar Zilsel, who, as early as 1918, published *Die Geniereligion*, 'The Religion of Genius,' in Vienna. According to Zilsel, this cult developed from the seventeenth century onwards and underwent further evolution during the Enlightenment and Romanticism, leading to the birth of true 'genius enthusiasts', such as the composer Richard Wagner, the man of letters Otto Weininger, the writer Houston Stewart Chamberlain—the latter very dear to the Nazis for his racial theories—and the Scottish writer Thomas Carlyle—author of the celebrated *On Heroes, Hero-Worship, & the Heroic in History* of 1841. Zilsel notes how contemporary culture is full of biographies of past geniuses, busts, effigies, and newspaper articles that do nothing but celebrate the great geniuses of Humanity. Geniuses to whom a salvific role is attributed, whose tombs are often the destination of pilgrimages, whose remains attract a devotion similar to that reserved for the relics of saints and martyrs. There is a notable absentee in Zilsel's work, Friedrich Nietzsche, whose thoughts on genius are actually quite complex. In Human, *All Too Human*, Nietzsche writes, for example, in aphorism 163: "Speak not of gifts, or innate talents! One can name all kinds of great men who were not very gifted. But they acquired greatness, became 'geniuses'...." Nevertheless, Nietzsche's insistence on the superior men of the future who will redeem the world and on the creation of new values undoubtedly went on to contribute—thanks also to the philosopher's posthumous fame—to the European cult of genius.

Around the same time, in February 1925, the well-known German neurologist Oskar Vogt, director and founder of the Kaiser Wilhelm Institute of Brain Research in Berlin, travelled to Moscow at the request of the Soviet government. Vogt's specialities—he was a true luminary of neurology—in fact, included genius brains. Not only that: Vogt was also a proponent of eugenics, and believed that his studies would contribute to the development and dissemination of superior brains. The aim of the trip was to verify, on neurological grounds—actually confirm—the genius of Vladimir Ilyich Lenin, who died on January 21 of that year. There were several reasons why Vogt was chosen, despite the fact that he was not Russian and was not a communist: he was a luminary, he sympathisd with the left, but, above all, he was a friend of Nikolai Alexandrovich Semashko, People's Commissar for Health and a supporter of eugenics. What attracted Vogt to Moscow was not only the possibility of studying Lenin's brain, but also that of setting up an institute to carry on the eugenic study of the genius. Russian scholars had obviously read European studies on eugenics with care, and dreamed. Russian politicians dreamed too: Leon Trotsky hoped that, one fine day, the average man would manifest capabilities comparable to those of Aristotle, Goethe and, why not, Marx. And said scholars included G. V. Segalin, a doctor and psychiatrist who proposed the creation of an 'Institute of Genius' to examine the bodies and brains—deceased, of course—of talented men. Inventor of his own discipline, 'ingeniology', Segalin believed he had also identified a 'creativity gene'. The enthusiastic ingeniologist also proposed establishing clinics to care for geniuses and child prodigies, treating the negative aspects of their condition and encouraging the positive ones.

But back to Vogt: the scholar immediately set to work with his wife Cécile, dissecting Lenin's brain, which was later divided into more than thirty thousand samples. In the meantime, Vogt began, together with Semashko, to prepare a draft for the creation of a research institute dedicated to the study of brain biology. Established and financed in 1926 through direct intervention by Stalin, the V. I. Lenin Institute for Brain Research was opened the following year. There was no shortage of work, so much so that the institute continued to work on Lenin's brain, as well as those of other well-known Russian revolutionaries and eminent figures. It also housed what would become known as the Pantheon of Brains, which brought together the brains—and other organs—of important Soviet public figures. And, on the subject of Lenin's brain, Vogt's studies confirmed what his devoted compatriots already knew, namely, his genius—the German scholar states, in particular, that Lenin was a "mental athlete."

As we have seen, nineteenth- and twentieth-century Europe was crisscrossed by a hunger for genius, by a cult for genius that sometimes took extreme forms, as in the case of the father of positivism, Auguste Comte, who sought to create a Religion of Humanity in which the cult of God was replaced by a cult for the great minds of history. But in Marxism, the story is different: the ideology in question is actually the antithesis of the glorification of the individual, so much so that Marx himself, in a review of Carlyle's work, specifically condemned the cult of genius. Moreover, there can be no peace between the eugenic approach advocated by Vogt and Marxist doctrine: eugenics and Marxism are, in fact, based on conflicting assumptions. Human characteristics would be hereditary for the former and socially constructed by education and environment for the latter. So much so that eugenics was eventually forced to yield to the ideological demands of the Communist Party, which ended up having it declared a bourgeois doctrine. 1930 saw the end of Vogt's Moscow institute, which came under the control of the Soviet Academy of Sciences.

Let us now return to Germany. Even Hitler, in the ominous *Mein Kampf*, has his say on genius, and, in particular, not only links it to the Aryan race, but considers it innate and believes its main catalyst to be war. Already, on April 17, 1920, the future Führer would say in a speech that Germany needed to be led by a genius, thus referring to a rhetoric of genius that was widespread at the time in German political discourse, not only among the people, but also in universities and among intellectuals. This rhetoric became even more insistent after the seizure of power by the Nazis, whose propaganda insisted on portraying Hitler as a genius, even having him appear in the company of the great geniuses of the past, appropriately "nazified." Films made specifically for the purpose contributed to the 'beatification' of German and non-German geniuses, and Hitler himself got a film treatment by Leni Riefenstahl, the infamous *Triumph of the Will*.

1.6   The Last Genius? The Democratisation of Genius

Meanwhile, on the other side of the Atlantic—and continuing after the Second World War—we witness the celebration of another genius, this time a real one: Albert Einstein. Einstein, the scientist who, thanks to his genius, was able to take a deep look at the intimate mechanisms that govern physical reality. In 1946, *Time* magazine crowned him a "cosmoclast"—i.e., "destroyer of the cosmos," thanks to his famous equation $E = mc^2$ that had led, in a somewhat circuitous way, to the atomic bomb and victory over Japan. Interestingly, in relation to this fame as a genius, Einstein was both pleased and annoyed—as he claimed that it interfered with his scientific work. And in the case of Einstein, it is also to be considered that the status as genius was awarded to him as a result of developing theories that the general public did not have the tools to understand.

As McMahon notes, Einstein produced a change of perspective in the public conception of genius: whereas, before him, genius was attributed mainly to artists, men of letters, and so on, it was now scientists who were taking centre stage. Could the brain of the world's most famous scientist have escaped science? Certainly not, and, in fact, on April 18, 1955, it was taken by an anatomist from Princeton Hospital, Thomas Harvey, who provided samples to a group of Californian neuroanatomists. In 1985, a study was published claiming that Einstein's brain contained a larger than average number of glial cells, cells that support neurons in the process of neurotransmission. In 1999, the *Lancet* published another study showing that Einstein's parietal lobe, the area responsible for linguistic, spatial and mathematical understanding, was fifteen per cent larger than average. If you want to get a personal insight into Einstein's brain, know that, upon Harvey's death in 2007, the brain was digitised and made available to download for as little as ten dollars—a total of almost three hundred and fifty slides, viewable on your computer.[3] As early as 1958, the philosopher Hannah Arendt introduced the theme of the popularisation of the concept of genius, but this process, according to McMahon, began even earlier, namely, immediately after the end of the Second World War, and led to the progressive democratisation of this concept. Newspapers and magazines celebrate this or that genius, and today, this label is attributed to a number of successful men, from Elon Musk to Jeff Bezos, to Steve Jobs, to Mark Zuckerberg, but also to figures from sports, fashion designers, and professionals of all kinds. And so, we are, perhaps, at a point in our cultural history where the line between genius and celebrity has become very thin, almost imperceptible.

And democratisation of genius also means commercialisation; just enter any bookshop and you will find—in the self-help section—countless volumes explaining how to become a genius. So, let's take a look at the most popular titles, many of which—we confess—even we had read, in the vain hope of improving our skills. Christine Ranck Christopher and Lee Nutter urge us to *Ignite the Genius Within*; Robert and Michele Root-Bernstien, in *Sparks of Genius*, illustrate *The Thirteen Thinking Tools of the World's Most Creative People*; in *Uncommon Genius*, Denise Shekerjian explains *How Great Ideas Are Born*; Kim Addonizio is more specific, dedicating *Ordinary Genius: A Guide for the Poet Within* to poets. We also mention *Thought Revolution—How to Unlock Your Inner Genius* by William A. Donius. Michael Gelb offers us two books, *Discover Your Genius: How to Think Like History's Ten Most Revolutionary Minds* and *How to Think Like Leonardo da Vinci: Seven Steps to Genius Every Day*. Our list also includes *Accidental Genius: Using Writing to Generate Your Best Ideas, Insight, and Content*, by Mark Levy; *Cracking Creativity: The Secrets of Creative Genius*, by Michael Michalko; *The Secret Principles of Genius: The Key to Unlocking Your Hidden Genius Potential*, by I. C. Robledo; *The Hidden Habits of Genius: Beyond Talent, IQ, and Grit-Unlocking the Secrets of Greatness*, by Craig M. Wright; and *Practical Genius: A 5-Step Plan to Turn Your Talent and Passion into Success*, by Gina A. Rudan and Kevin Carroll. Finally, a note of encouragement is offered by David Shenk in *The Genius in All of Us: New Insights into Genetics, Talent, and IQ*.

If everyone can potentially be a genius, asks McMahon, doesn't that mean that no one is really a genius? Does the democratisation of genius coincide with the end of the figure of the genius, and also with the end of the cult dedicated to it? Was Einstein the last genius worthy of the name? These are legitimate questions, given the collaborative, and therefore anti-individualistic, nature of much modern scientific enterprise. McMahon then

---

[3] https://itunes.apple.com/us/app/nmhm-anatlab-harvey/id555722456.

continues along a path that we would rather not take, namely, that of post-structuralist analysis that rejects the role of the subject and instead emphasises the social—and therefore collective—nature of discoveries, inventions and works of art. Despite all of this, the subject of genius continues to fascinate the general public, and they're not alone. One thinks, for example, of American entrepreneur Robert K. Graham's initiative—dating back to the 1980s and 1990s—to set up the Repository of Germinal Choice, an attempt to create a sperm bank of Nobel Prize winners. Despite the bank being funded by William Shockeley—remember him? Termites' excluded boy—and other Nobel Laureates, it closed its doors in 1997. Also deserving of special mention is Allan Snyder, a scholar at the University of Sydney who tried to develop a 'creativity cap,' a device that he claims can stimulate the same abilities held by the Savants in everyone.

And let us not forget Artificial Intelligence research, which aims, among other things, to reproduce human capabilities in machines, as is the case with Deep Blue, which managed to beat chess master Garry Kasparov in 1997. For now, attempts to imitate creativity and genius in machines are still vague.

Notwithstanding McMahon's dissent, we believe that the discourse on genius should not be framed solely within a sociological and anthropological context, but also within a context of individual psychology. In our humble opinion, the subject of genius is still relevant today, and can be framed within a scientific context that preserves the individual. A psychology of genius, in short, to which we will devote the rest of this book.

2

Genius and Its Siblings

2.1 Think

It is inevitable: when we speak of genius, we end up referring to other
psychological constructs, which run parallel to it and whose study is often
confused with that of human eminence. This chapter will be devoted to them.
In particular, we will deal with intelligence, creativity, wisdom, talent and
charisma, five 'things' that have provoked much debate and research over the
past decades. We will try to understand what these five attributes consist of,
and see how they intersect with that of genius.

Let us start, then, with intelligence, a construct that is much less
easy to define than is commonly thought—in this regard, Edwin Boring's
famous reductionist and somewhat provocative definition comes to mind,
in which intelligence is precisely what is measured by intelligence tests.[1]
For example, a famous symposium was held in the USA in 1921, entitled 'Intelligence and Its Measurement,' whose participants—including the
aforementioned Terman—offered a long series of definitions of intelligence.[2]
Edward Thorndike, one of the leading exponents of behaviourism, defines
intelligence as "the power of good responses from the point of view of truth
or facts," while Terman's definition refers to "the ability to carry on abstract
thinking." Other definitions include "sensory capacity, capacity for perceptual

---

[1] Boring, Edwin G., *Intelligence as the Tests Test It*, New Republic 36, 35–37, 1923.
[2] Various Authors, *Intelligence and Its Measurement: A Symposium*, Journal of Educational Psychology,
12(3), 123–147, 1921.

© The Author(s), under exclusive license to Springer Nature
Switzerland AG 2023
R. Manzocco, *Genius*, Springer Praxis Books,
https://doi.org/10.1007/978-3-031-27092-5_2

recognition, quickness, range or flexibility of association, facility and imagination, span of attention, quickness or alertness in response"; "having learned or ability to learn to adjust oneself to the environment"; "ability to adapt oneself adequately to relatively new situations in life"; "the capacity for knowledge and knowledge possessed"; "a biological mechanism by which the effects of a complexity of stimuli are brought together and given a somewhat unified effect in behaviour"; "the capacity to inhibit an instinctive adjustment, the capacity to redefine the inhibited instinctive adjustment in the light of imaginally experienced trial and error, and the capacity to realise the modified instinctive adjustment in overt behaviour to the advantage of the individual as a social animal"; "the capacity to acquire capacity"; "the capacity to learn or to profit from experience."

The starting point of the scientific study of human intelligence is the basic idea—obvious to some, not to others—that human beings would be characterised by individual differences, that people would possess this or that quality to a greater or lesser extent than their fellow human beings. In short, that human beings would not be equal. And this would also apply to intelligence, of course, to be understood as the general ability to reason, abstract, plan, become aware, and so on—this as far as the definition in our natural language is concerned.

It all begins with two characters we have already come, at least in part, to know, Sir Francis Galton and Alfred Binet, who not only kick-started this type of study, but also established the general terms of the debate, which continues to this day. Beware, however: Galton offers us a very different definition of intelligence from the one in vogue today. For the scientist, there are two factors that characterise individual intelligence, namely, energy—i.e., the ability to withstand fatigue, particularly mental fatigue—and sensitivity—i.e., the ability to discriminate sensory data, such as brightness, colours, smells, and so on, with finesse. Not content with theoretical work, for seven years, from 1884 to 1890, Galton ran a laboratory in London in which he precisely measured the ability of a certain number of volunteers to discriminate sensory stimuli in a more or less fine way—for example, the ability to distinguish with closed eyes the slightly different weights of two apparently identical objects. Note the irony of all this. Starting with Darwin, Darwinism has branched out in different directions, and according to some interpretations, it would be possible to hierarchically order the different living forms according to their complexity. From this point of view, the human being would be considered the most complex creature; Galton, promoter of a similar view, would not hesitate to consider humans as more evolved than dogs, but according to the approach he advocated in the field of intelligence, our faithful four-legged friends would in certain respects—for example, olfactory discrimination—certainly be superior to us.

Binet thinks differently. We have already mentioned the origins of his IQ test, and now we add some important details. We are particularly interested in the concept of intelligence introduced by the scholar. It is based not on perceptual discrimination, but on judgment, i.e., the ability to formulate a thought about a certain problem. For Binet, Galton's concept is meaningless, and in this regard, the French scholar cites the famous case of Helen Keller (1880–1968), an American writer and activist who was deprived of sight and hearing from an early age. Obviously, her intellectual level would be zero according to Galton's criteria.

There are three elements that make up intelligence, namely, direction—knowing what goal or result one wants to achieve—adaptation—the ability to monitor thought operations as they occur and adapt them—and control—the ability to criticise one's thoughts and actions. For Binet, there are two types of intelligence: conceptual intelligence—based on words and ideas, as in the case of logical reasoning—and instinctive intelligence—based on feelings.

With his approach, Binet can be said to have hit the nail on the head, so much so that the Stanford-Binet Intelligence Scale derives directly from the French scholar's work, while the other available tests—the Wechsler tests, developed in the 1950s by David Wechsler—are also indebted to him. One aspect of Binet's work that has rather been ignored is the idea that human intelligence is modifiable, and could therefore be enhanced. That is, for the scholar, it is possible in principle to become more intelligent—an idea that is still fairly controversial to this day.

A third character, Charles Spearman, also enters our story. Living at the turn of the nineteenth and twentieth centuries, the Englishman Spearman proposes a theory of intelligence that is still very popular, and based on two factors. In practice, human intellectual capacities would depend on a general factor 'g' common to all of the mental tasks we perform, and a specific factor 's'—or rather, a set of factors—linked to the specific task we set out to perform. It is not clear, according to Spearman, what this 'g' consists of, but it could be a generic 'mental energy'. Critical of the theories in question, the British psychologist Godfrey Thomson argued, in 1939, that the 'g' factor did not lie upon a single psychological construct, but a very broad set of factors, which he called 'bonds'.

The British-American psychologist Raymond Cattell then introduces a widely accepted distinction in psychology, that between 'fluid ability' and 'crystallised ability'. Basically, for him, there would be a hierarchy of abilities, with general 'g' ability at the top, and fluid ability—the ability to reason abstractly, measured by mathematical and analogical tests—and crystallised ability—consisting of knowledge accumulated over a lifetime, and measured by tests related to personal culture and available vocabulary—immediately below.

In contrast, John B. Carroll, an American, proposed, in the 1990s, a 'three-layered theory,' which sees 'g' at the top, fluid intelligence, crystallised intelligence and other capacities—memory, speed of execution, and so on—in the middle, and a wide range of very specific abilities underneath.

It is worth mentioning at this point the Swiss Jean Piaget, who, in his complex system of thought concerning the nature of the human mind and learning, distinguishes between assimilation—i.e., the absorption of new information and its arrangement in the cognitive structure—and accommodation—i.e., the transformation of existing cognitive structures in the light of new experiences that are not easily assimilated. Taken together, accommodation and assimilation constitute what Piaget calls equilibration, i.e., the process of organising sensory information. Extremely influential, Piagetian theory still lives on today, despite the inevitable modifications and criticisms it has undergone over the years.

If, for Piaget, biology is the basis of intelligence, figures such as Lev Vygotsky and Reuven Feuerstein—the latter actually also influenced by Piagetian thought—have instead pointed the finger at the environment in which the mind develops. For Vygotsky in particular, the basis of intelligence is a process he calls 'internalisation,' that is, the transport, or rather the reconstruction, within the interiority of the individual of an operation observed outside. And so, what we see on the outside becomes part of us. For example, abstract thought, which would represent a form of internalised language. Feuerstein, for his part, developed a very interesting approach, that of 'cognitive modifiability,' whereby intellectual capacities could be increased through exercise and, in particular, through a series of tools developed by the scholar himself over the years; in essence, the scholar subdivided human cognitive capacities into a series of very specific tasks, upon which it is possible to practise through the exercises he developed.

Finally, we cannot fail to mention, albeit in passing, the attempts to explain human intelligence in cognitive terms—e.g., those of George Miller, Eugene Galanter and Karl Pribram, the founders of this field in the 1960s—interpreting it, for example, as a process of information processing analogous to that carried out by computers.

There has, of course, been no lack of neurological approaches, which have attempted to root intelligence in the biological processes that take place in our brains. For example, in 1949, the Canadian Donald Hebb distinguished among Intelligence A—that is, the possession of a good brain and a good brain metabolism, two purely physical factors, Intelligence B—the functioning of the brain at full power, that is, when the developmental phase is over—and Intelligence C—which consists of the result obtained by taking an IQ test.

An attempt to root intelligence in observable neural activity was made in the 1970s by Russian psychologist Alexander Luria, for whom different areas of the brain would correspond to different functions and thoughts. In particular, the brain would consist of three units, the arousal unit, the sensory input unit and the unit responsible for organisation and planning—corresponding to the frontal cortex.

And we have to go back to 1836, to the work of a little-known French country doctor, Marc Dax, to find the first analysis of so-called hemispheric specialisation. Indeed, as is well known, the brain consists of two specialised hemispheres, the right and the left, studied first by Paul Broca and then by the physiologist and Nobel Prize winner Roger Sperry. In the 1960s, the latter suggested that they often behave, in all respects, as two separate brains, that the left hemisphere would be in charge of analytical and verbal functions, while the right would deal with imagery and synthesis. Today, we know that this view, despite having several points in its favour, is a simplification, and that the two hemispheres work closely together.

More recently, attempts have been made to link intelligence to specific factors of neural functioning, e.g., the speed of neuronal conduction, whether central—in the brain—or peripheral—e.g., in the limbs. In essence, the faster a subject is neuronally, the more intelligent he or she is. Alternately, there have been studies of glucose metabolism in this or that area of the brain, with the idea that those with a more efficient brain metabolism become less easily fatigued, and therefore are or would be more intelligent. Seemingly elegant in their simplicity, these solutions do not, however, clarify their causal direction, i.e., to put it simply, whether cognitive abilities depend on neuronal speed, for example, or whether it is the latter that depends on the former. This being neurology, a reference to brain size could not be far off, and, indeed, here it is: in 1991, Lee Willerman, Robert Schultz, Neal Rutledge and Erin Bigler in particular argued for the existence of a correlation between larger brains and the brain's ability to develop more complex synaptic connections.[3] However, this approach still remains controversial.

---

[3] Willerman, Lee; Schultz Robert; Rutledge, Neal, and Bigler, Erin, In vivo *brain size and intelligence*, 'Jntelligence,' Volume 15, Issue 2, Pages 223–228, April–June 1991.

And, in the age of genetics, there are obviously also those who have sought the genetic basis of intelligence. This is a complex issue, punctuated by various attempts to separate the influence of genes from that of the environment, and, in principle, we say that the literature in question recognises heredity as playing a not insignificant role in human intellectual abilities.

Let us now turn to systemic approaches to human intelligence, some of which have become very popular. These are models that attempt to break free from the strictures of IQ and develop a global view of intelligence. In particular, we mention the work of Howard Gardner, the well-known theorist of Multiple Intelligences. According to this model, developed since the 1980s, there are eight—and perhaps more—intelligences, namely, the linguistic, the logical-mathematical, the musical, the bodily, the spatial, the intrapersonal—the famous ability to 'look inside oneself'—and interpersonal, the naturalistic, and, finally—perhaps—the existential and spiritual. The big problem with this approach is that, although it is quite fascinating, it is currently lacking strong empirical confirmation. It is a psychological construct in need of extensive experimentation.

Another psychological construct, that of 'emotional intelligence', has been widely discussed in the media over the years. Resulting from the work of various authors, this idea was popularised, in particular, by Daniel Goleman in the mid-1990s. It consists of the ability to perceive, evaluate and regulate emotions, a skill that is undoubtedly very useful in various areas of life.

Let us close the review with Robert Sternberg's triarchical approach, which sees human mental capabilities as a global system composed of three factors, namely, intelligence—the analytical aspect, dedicated to analysing what is already known-, creativity—the innovative aspect, consisting of the ability to venture into unexplored territories—and wisdom—understood as practical intelligence aimed, however, at the common good.[4] Here, then, is a general overview of theories on intelligence. A quality that does not, however, seem to be linked too strongly to genius, as we are about to see.

2.2   Create

Let us now turn to creativity, which is the human trait most closely linked to the theme of genius, to the extent that research on the former overlaps in many places with reflection on the latter. Let us first give an operational definition of creativity, to be understood as the capacity to produce works that

---

[4] Sternberg, Robert J., *Wisdom, Intelligence, and Creativity Synthesized*, Cambridge University Press, New York 2003.

are new—that is, original, surprising—and useful for something, that is, in which the social context recognises value, even if only as a source of pleasure or wonder. Even this last attribute—that of context-dependent usefulness—is, indeed, problematic, but this is an issue we will explore later.

For a long time, psychology ignored this topic, and the speech in 1950 by the then-president of the American Psychological Association, J. Paul Guilford, is famous in this respect. The scholar declared, in fact, that psychological research in the field of creativity was minimal—only two-tenths of one per cent of the studies published at the time were devoted to the subject. After Guilford's lecture, things improved a little, but the subject of creativity remained secondary until relatively recently.

Let us therefore take a look at the main approaches used towards the study of creativity, without forgetting that, from the very beginning—that is, since antiquity—this subject has been characterised, just like genius, by a mystical aura. The first attempts to explain this phenomenon, in fact, refer to divine intervention, to the creative person understood as a simple vessel of superior forces of non-human origin. In particular, we remember Plato, according to whom a poet does nothing more than declaim what the Muses suggest to him. And even in times closer to us, prominent authors refer to Muses and 'demons' in the sense that we have already illustrated above. For example, Rudyard Kipling speaks of a 'Daemon' who dwells in his pen and whom he must obey.[5] And even in common thinking, that of the person in the street, this idea that inspiration comes 'from outside' is widespread. Obviously, this is a view that, for the purposes of psychological research, is completely useless and fruitless, since it actually constitutes a renunciation of explanation. And while, on the one hand, we have this mystical, literally millenarian attitude, on the other, we have witnessed, in recent decades, the rise of another kind of approach that is different, but equally 'harmful'—if one can call it that—to the scientific understanding of creativity. We are talking about the pragmatist-commercial approach, characterised by authors, books and courses that promise to everyone the development of exceptional creative abilities. Not that this is a completely incorrect approach, quite the contrary: in fact, one often finds good advice in the books in question. However, it tends to propose a simplified version of the creative process. We refer, for example, to the—very famous—work of Edward De Bono, whose reflections on so-called 'lateral thinking' have been a huge commercial success. The author proposes a series of conceptual tools that he says should stimulate creativity, such as the famous "thinking hats." Not that these things are not useful, again, quite

---

[5] Kipling, Rudyard, *Something of Myself*, R. & R. Clark, Edinburgh 1937, p. 162.

the contrary. The problem is that they lack a solid underlying psychological theory that would allow us to understand how the creative act actually works.

In between these two extremes lies psychology tout-court, with the potential solutions proposed by academic researchers. Among the schools of thought that have contributed to the study of creativity is the psychodynamic approach, i.e., psychoanalysis and the schools of thought derived from it. Already, Freud had tried to explain human creativity by way of the unconscious; in particular, the creative act would represent a way to express unconscious desires in a manner that is socially and publicly acceptable. The father of psychoanalysis cites the case of eminent creators, such as Leonardo Da Vinci, to support his view of things.

Other epigones of psychoanalysis would deal with creativity. For example, in 1958, the American psychoanalyst Lawrence Schlesinger Kubie introduced—in his book *Neurotic Distortion of the Creative Process*—the idea that the source of creativity was the 'preconscious,' a psychic dimension situated between the true unconscious and conscious reality. In it, thoughts would be relatively free of constraints but interpretable—unlike the actual unconscious, the exploration of which requires lengthy psychoanalytic work. Contrary to what others had claimed, for Kubie, not only would unconscious conflicts not be the source of creativity, they would represent an obstacle, as they would lead the subject to fixate on repetitive thoughts. More generally, psychoanalysis, over time, has come to support the importance for creativity of both primary processes—ego-related, 'infantile,' dreamlike—and secondary processes—logical-rational and verbal. Having said this, however, let us emphasise how neither psychoanalysis nor the other psychological schools of thought of the first half of the twentieth century—for instance, Gestalt, which dealt specifically only with the notion of insight—treated creativity with due attention.

Let us return, then, to Guilford's lecture, in which, among other things, the scholar takes note of an essential problem of any systematic study of creativity, namely, the difficulty we would encounter in subjecting a sufficient number of highly creative characters to laboratory tests. After all, it is not every day that one meets a Picasso or a Van Gogh, and an even rarer day in which one would be able to count on their regular collaboration. Creativity, however, is not only the prerogative of those who have a lot of it, but also of all those who want to try their hand at such work, which is why Guilford proposed that such people study it on their own. in generic subjects using paper-and-pencil tests developed for the occasion. And so, in 1974, the American psychologist Ellis Paul Torrance developed a special test to analyse creativity in an everyday setting. Composed of very simple verbal and figural tasks, the Torrance Test of Creative Thinking assesses creative qualities such as fluency—how many pertinent answers one is able to give—originality—i.e., the statistical frequency of the answers given, in other words, their rarity—flexibility—how many different categories the answers given fall into—and richness of detail.

At this point, one might ask whether there is any relationship between this type of test and the classic IQ test. Already, in 1926, Chaterine Cox was convinced that there was a direct proportionality relationship between intelligence and creativity: in practice, according to the scholar, highly creative people would also be remarkably intelligent. And her mentor Terman emphasises the centrality of Galton's IQ—estimated at around 200 points—to his intellectual eminence. Galton, Cox tells us, recognised capital letters at twelve months, and the whole alphabet, upper and lower case, at eighteen months; he could make his signature before the age of three. By the age of four, our child prodigy knew Latin grammar and read a bit of French, by five he was engrossed in the Iliad and the Odyssey, and by seven he was dabbling in Shakespeare. The scholar's conclusion is that a high intelligence, though not necessarily very high, connected with high persistence—very high in this case—led to creative eminence.

Subsequent research has highlighted the peculiar relationship between intelligence and creativity, a fact that is of particular interest to us in our discussion of genius. In particular, the US psychologist Joseph Renzulli tells us that creative people tend to have an above-average IQ, often exceeding 120 points.[6] So far, so good: in order to be creative—and therefore to aspire to the licence of genius—one must possess a certain degree of intelligence. That 120 points is not exactly a rigid dividing line, so that it allows us to derive other averages: for example, Cox's geniuses have an average IQ of 165. What is important is that, once the minimum level of 120 points is reached, the IQ no longer seems to count. In short, the level of creativity and the level of intelligence seem to go hand in hand up to 120 IQ points, after which they progressively decouple from each other, so that a high IQ does not necessarily mean a high level of creativity, and vice versa. This was highlighted by the work of, among others, American psychologist Frank Barron, who examined a number of architects.[7] In the study in question—carried out on subjects

---

[6] Renzulli, Joseph S., *The Three-Ring Conception of Giftedness: A Developmental Model for Creative Productivity*, In: Sternberg, Robert J. and Davidson, Janet E., Eds., *Conceptions of Giftedness*, Cambridge University Press, New York, 53–92, 1986.

[7] Barron, Frank, *Creative person and creative process*, Holt, Rinehart & Winston, New York 1969 p. 42.

whose average IQ was 130—the relationship between creativity and intelligence was statistically insignificant. And, according to some scholars, such as Dean Keith Simonton, a very high IQ could even interfere with creativity.[8] Having a high IQ leads possessors to receive very positive feedback for their analytical skills, which might discourage them from developing their creative abilities.

The preceding reflections should not, however, lead us to think that there is no correlation between intelligence and creativity; rather, it oscillates between weak and moderate, but, above all, it will vary depending on the discipline we are considering. Fields such as mathematics will, in fact, require greater analytical skills than those required by art, for example.

Undoubtedly, the psychometric approach illustrated here—which has produced various types of tests over the years—has made research easier; it is not without its criticisms, however. For instance, its excessive simplicity—i.e., to really test creativity, one would need exercises that are more articulated than those generally proposed in tests of this type—or the fact that it precisely targets ordinary people—i.e., it does not really allow one to explain creative eminence, that which, in practice, differentiates us ordinary mortals from the great creative people.

Alongside the cognitive approach, others have developed: in particular, the psychosocial approach, which, instead of dealing with the creative process itself, prefers to focus on the psychological traits of creative people, their motivations, and so on. This has been dealt with, for example, by Teresa Amabile,[9] Hans Eysenck,[10] Frank Barron,[11] and many others. These are often correlation studies, which highlight how creative people tend to show a number of common traits, such as a tendency to take risks, a search for and attraction to complexity, self-confidence and a strong independence of judgment. Amabile and Barron went on to emphasise the importance of intrinsic motivation—'art for art's sake,' we might say—over extrinsic motivation—i.e., social rewards and whatnot. Also, in the field of personality psychology, we cite the classics by Abraham Maslow[12] and Carl Rogers[13]—exponents of

---

[8] Cf. Simonton, Dean K., *Greatness: Who makes history and why?* Guilford, New York 1994.

[9] Amabile, Teresa M., *The social psychology of creativity*, Springer, New York 1983.

[10] Eysenck, Hans J., *Creativity and personality: A theoretical perspective.* Psychological Inquiry, 4, pp. 147–178, 1993.

[11] Barron, Frank, Op. Cit. 1969.

[12] Maslow, Abraham, *Toward a psychology of being,* New York: Van Nostrand, New York 1968. The scholar distinguishes between primary creativity—that which the person uses to seek self-fulfilment—and secondary creativity—that which is specifically studied by psychology and concerns contributions made in this or that field of science and art.

[13] Rogers, Carl R., *Toward a theory of creativity,* ETC: A Review of General Semantics, 11, 249–260, 1954.

the so-called 'third force,' i.e., humanistic psychology—who, in the course of their work, always emphasised the importance of factors such as spontaneity, creativity and freedom in self-fulfilment. The latter in particular emphasises the need for an environment free of judgment for human potential to be realised. The evolutionary approach to creativity—promoted by Donald Campbell, Dean Keith Simonton and, more recently, Gary Cziko[14]—will be discussed in detail later. For now, let us just say that this model holds that at the heart of creativity is a process of random idea generation, whereby great creators in science and art would first of all have a lot of ideas, which would increase the likelihood of having brilliant ideas. Not everyone agrees with this model; on the contrary: scholars such as Robert Sternberg argue, for example, that creative people may have more or fewer ideas than non-creative people, but what matters above all is the fact that the former have *better* ideas than the latter. That is, the idea that genius ideas are simply the result of randomness seems implausible to many, and the question is still open. Additionally, we will take a closer look at Mihaly Csikszentmihalyi's systemic approach later on,[15] while also mentioning in passing the work of Howard Gardner, who has decided to apply his multiple intelligences approach to the field of creativity, dedicating a special volume to the topic.[16] In it, the scholar examines the lives of seven eminent personalities, each of whom excelled in one of the intelligences he studied, so we have intrapersonal intelligence and Sigmund Freud, spatial intelligence and Pablo Picasso, logical-mathematical intelligence and Albert Einstein, bodily-kinesthetic intelligence and Martha Graham, musical intelligence and Igor Sravinsky, and interpersonal intelligence and Mohandas Gandhi. Gardner points out, however, that all these people excelled in *several* intelligences, not just one. Two other aspects that the scholar emphasises are: renunciation, i.e., all of these people would have had to give up the everyday pleasures of life in order to achieve a creative leadership role; and support, i.e., they would have had a supportive social network that would have enabled them to devote themselves to their own endeavours.

And finally, *dulcis in fundo*, we cite Robert Sternberg's approach, according to which the creative process is the result of a decision-making process, in

---

[14] Cziko, Gary A., *From blind to creative: In defence of Donald Campbell's selectionist theory of human creativity*, Journal of Creative Behavior, 32, 192–208, 1998.

[15] Csikszentmihalyi, Mihaly, *Creativity: Flow and the psychology of discovery and invention*, Harper-Collins, New York 1996.

[16] Gardner, Howard, Creating *Minds: An Anatomy of Creativity Seen Through the Lives of Freud, Einstein, Picasso, Stravinsky, Eliot, Graham, and Ghandi*, Basic Books, New York 2011.

other words, the decision to be creative.[17] In practice, this theory—christened Investment Theory by the scholar—consists of the idea that "creative people decide to buy low and sell high in the world of ideas—that is, they generate ideas that tend to 'defy the crowd' (buy low), and then, when they have persuaded many people, they sell high, meaning they move on to the next unpopular idea."[18] The scholar then introduces a rather important distinction, namely, between what he calls 'creativity with a lower-case c' and 'creativity with a capital C,' defined as—in the first case—original contribution only with respect to oneself or—in the second case—original contribution with respect to a certain field. Creativity also constitutes, in a metaphorical sense, a form of propulsion, i.e., the creator aims to propel a certain field from one certain point in the space of possibilities to another, and to lead their colleagues there as well; to create therefore also means, in a certain sense, to exercise leadership.

2.3   Judging

Let us return to intelligence for a moment: it is indeed interesting to take a look at ways of conceiving intelligence other than those found in the West. In this regard, Sternberg has carried out an interesting cross-cultural analysis, particularly in relation to China.[19] According to Confucian thought, intelligence would be particularly associated with benevolence and behaving justly, while Taoist thought emphasises humility and in-depth knowledge of oneself and one's surroundings. Which brings us finally to the connection between intelligence and wisdom, another psychological construct, which Sternberg defines—dictionary in hand—as the "power of judging rightly and following the soundest course of action, based on knowledge, experience, understanding."[20] How do we study wisdom? In various ways: first, we have philosophy itself, which has always dealt with wisdom; then, we have psychology, in its two approaches, the implicit-theoretical and the explicit-theoretical. To address wisdom from the perspective of philosophy would take us too far, so we will focus on what psychological research has to

---

[17] Sternberg, Robert J., & Lubart, Todd I., *Defying the crowd: Cultivating creativity in a culture of conformity*, Free Press; New York 1995.

[18] Sternberg, Robert J., *Wisdom, Intelligence, and Creativity Synthesized*, Cambridge University Press 2003, New York, p. 106.

[19] Yang, Shih-ying & Sternberg, Robert J., *Conceptions of intelligence in ancient Chinese philosophy*, Journal of Theoretical and Philosophical Psychology, 17(2), pp. 101–119, 1997.

[20] Sternberg, Robert, op. cit. p. 147.

say about it. In particular, the implicit-theoretical approach seeks to understand not what 'true wisdom is,' but what the definition of wisdom implicit
in common perception is, whether right or wrong. According, in particular, to the research of two psychologists, Stephen Holliday and Michael
Chandler—who studied about five hundred people—wisdom in common
perception would include six ingredients, namely, the ability to reason, the
ability to learn from the environment, judgement, quickness in the use of
information, perspicacity, and sagacity; thus, wisdom and intelligence seem
to be connected in this case.[21] Let us now turn to the explicit-theoretical
approaches, all characterised by a strong, i.e., formalised, theory of wisdom.
With regard to such approaches, the German psychologist Paul B. Baltes,
who has incorporated his own discourse on wisdom into his broader research
programme devoted to aging, mental abilities and the life cycle, is certainly
worth mentioning.[22] According to the German scholar, wisdom is composed
of five elements, namely, an in-depth knowledge of human life and its case
histories, procedural knowledge—i.e., on the strategic course of action with
regard to this or that case in life—the ability to contextualise life events and
relationships according to the phase of life in which they occur—the ability to
relativise values and points of view—and finally, an awareness of the fundamental unpredictability of life and the ways in which to relate to this fact.
In the course of their research, Baltes and colleagues highlighted how intelligence and wisdom tend to overlap but are not identical[23]; how elderly people
referred to by others as wise were able to perform tasks related to 'wise' evalua-
tions with the same skill as clinical psychologists, and how people up to eighty
years in age were able to perform the aforementioned tasks with 'youthful'
efficiency.[24] And there are also those who have tried to call Piaget into question, considering wisdom as a further stage in human cognitive development.
Wise people would thus be able to think dialectically and reflectively, with
the understanding that truth is not always absolute.[25] For other theorists,

[21] Holliday, Stephen G., & Chandler, Michael J., *Wisdom: explorations in adult competence*, Karger, Basel 1986.

[22] Baltes, Paul B., & Smith, Jacqui, *Toward a psychology of wisdom and its ontogenesis,* In R. J. Sternberg (Ed.), Wisdom*: Its nature, origins, and development,* Cambridge University Press, New York 1990, pp. 87–120.

[23] Staudinger, Ursula M., Lopez, David F., & Baltes, Paul B., *The psychometric location of wisdom-related performance: Intelligence, personality, and more?* Personality & Social Psychology Bulletin, 23, 1997, pp.1200–1214.

[24] Baltes, Paul B., Staudinger, Ursula M., Maercker, Andreas & Smith, Jacqui, *People nominated as wise: A comparative study of wisdom-related knowledge,* Psychology and Aging, 10, 1995, pp. 155–166.

[25] See, for example, Kitchener, Karen S. & Brenner, Helene G., Wisdom *and reflective judgment: Knowing in the face of uncertainty,* in: Sternberg Robert J. (Ed.), *Wisdom: Its nature, origins, and development,* Cambridge University Press, New York 1990, pp. 212–229.

then, wisdom would consist not so much in the ability to solve important problems as in the ability to identify them.[26] Also interesting is the approach of Mihaly Csikszentmihalyi and Kevin Rathunde, who have tried to introduce an evolutionary interpretation of wisdom. The vast majority of wisdom theories are, in fact, ontogenetic—i.e., they try to explain how a single individual develops this trait, step by step—while that of these two scholars is, as they say in biology, phylogenetic, i.e., it traces its origin through the ages.[27] In particular, according to Csikszentmihalyi and Rathunde, the trait of wisdom must be the result of selective pressures, particularly with regard to three aspects, namely, the cognitive—wisdom as a specific ability to acquire information-, the social—wisdom as a socially valued mode of behaviour—and the psychological—wisdom as a personality trait that has been deemed desirable.

And finally, we come to Sternberg's own approach, which sees wisdom as the application of intelligence and creativity in order to achieve collective balance and the common good—from the perspective of commonly accepted values. In particular, the balance concerns individual and collective interests in the medium and long term, with the aim of adapting to the environment, changing the environment or selecting a new environment. Moreover, wisdom is indeed based on knowledge, but tacit knowledge, which is learned through life experiences and not through study.

2.4   Having Talent

As far as the notion of talent is concerned, it originally indicated a unit of weight measurement—especially in relation to gold and silver, hence economic value—introduced in Mesopotamia at the end of the fourth millennium BC and subsequently subject to several variations—in New Testament times, it corresponded to 58.9 kg. In the sacred text in question, the aforementioned term is used in the parable of the same name (Matthew 25:14–30); later, in the West, the term—also as a result of the interpretation of the parable in question—ends up indicating a natural ability or inclination. Etymologically, the term 'talent' derives from the Latin 'talentum,' which, in turn, derives from the Greek 'talanton,' meaning 'balance' or 'sum.' In

---

[26] See Arlin, Patricia Kennedy, *Wisdom: the art of problem finding*, in: Sternberg Robert J. (Ed.), *Wisdom: Its nature, origins, and development*, Cambridge University Press, New York 1990, pp. 230–243.

[27] Csikszentmihalyi, Mihaly, & Rathunde, Kevin, *The psychology of wisdom: An evolutionary interpretation*, in: Sternberg Robert J. (Ed.), *Wisdom: Its nature, origins, and development*, Cambridge University Press, New York 1990, pp. 25–51.

short, we have moved from economic/monetary value to performance value. Today's definition of talent primarily indicates artistic and scientific excellence, but also sporting excellence or other areas of life—a person may have, for example, a 'talent for politics,' a 'talent for business,' and so on. The implicit idea here is that talent—just like in the Gospel parable—is 'buried' and must be brought out through hard work over time.

As far as the relationship between talent and genius is concerned, it all starts with the work of Lewis Terman, who attempted to study genius by recruiting people who excelled in 'talent areas' such as music, art, mechanics, and so on. Terman therefore originally did not rely on IQ, but he quickly decided to abandon the talent-based approach due to the objective difficulty in establishing different 'degrees of talent,' i.e., distinguishing between more and less talented people. Moreover, the talented children he studied showed him that they also had a higher than average IQ, so Terman decided to rely on the latter for his own research. Around the same time, the American psychologist Leta Stetter Hollingworth—the creator of the concepts of 'gifted' and 'giftedness'—distinguished between genius and talent: the former would consist of the 'ability to achieve mental perfection,' whereas talent would be a remarkable ability, but one that does not reach the level of genius.[28] She later modified the latter definition, using the term talent to indicate a specialised aptitude. It would thus be similar to the high levels of the specific 's' factors theorised by Spearman, while giftedness would be comparable to the latter's general 'g' factor. In 1983, the US psychologist Abraham Tannenbaum proposed a 'social' subdivision of giftedness into four categories. The first category, 'scarcity talents,' are all those talents that are scarce but very useful to society—for example, all those people who contribute to society through innovations that make the world safer, healthier, more functional, and so on. The 'surplus talents' are all those talents desired by society but not indispensable, such as those displayed by musicians, actors, artists, and so on. Not that these contributions are useless, for Tannenbaum; simply less essential than contributions in the fields of medicine, food, and the like. In between these two categories, there are the 'quota talents,' people with in-demand skills who do not require any particular creativity, just a high level of performance, such as doctors, teachers, lawyers, engineers. Finally, there are the 'anomalous talents,' those exhibited by people with exceptional skills that are, however, of limited use and are more a source of amusement and amazement: extreme sports practitioners, speed readers, very skilled but anachronistic craftsmen.[29]

---

[28] See Hollingworth, Leta S., *Gifted children: Their nature and nurture*, Macmillan, New York 1926.
[29] See Tannenbaum, Abraham, *Gifted children's psychological and educational perspectives*, New York: Macmillan, New York 1983.

The Canadian psychologist Francoys Gagné distinguishes between giftedness and talent: giftedness corresponds to a high level of aptitude for something, while talent consists of this aptitude developed through intensive and sustained work. In turn, giftedness means an underlying potential, which may be high or low. Then, there are the catalysts; a person may be born with a gift for a certain discipline, but for this to develop into talent, an appropriate disposition, a certain environment and a certain education are required. For Gagné, there are two groups of catalysts: intrapersonal ones—motivation, temperament, and so on—and external ones—places, opportunities, people.[30] During the 1980s, the American psychologist Benjamin Bloom also dealt with talent; his research indicates that the development of talent, especially world-class talent, involves not only an enormous amount of effort and time, but also family support, rewards—i.e., experiences of success—and very sophisticated levels of teaching and coaching.[31]

2.5   Fascinate

And here, we finally come to charisma, a psychological quality often attributed to people of genius. Indeed, who has never found themselves fascinated by this or that genius, and his or her seemingly superhuman abilities? Charisma is a word that—pardon the pun—has been talked about a lot. It is a story about two thousand years long, one that parallels the entire history of the West and, in particular, its dominant religion, Christianity. In fact, as the historian of ideas John Potts tells us in his excellent *History of Charisma*, it was from the early Christian religion that the term 'charisma' emerged, only to disappear from the radar towards the end of the third century and then reappear in the modern era, becoming a term in common use—especially in politics, the media and the world of entertainment—in the second half of the twentieth century.[32] It was the German sociologist Max Weber who took up and, in a sense, 'secularised' the concept of charisma in modern times—at the turn of the nineteenth and twentieth centuries. The charisma of Weber is quite different from that promoted by St. Paul, and yet many later scholars have weakened the conceptual boundaries between the two definitions. For example, the various charismatic modes identified by St. Paul did not include leadership, which is at the heart of Weber's definition. And so,

[30] Gagné, Francoys, *From gifts to talents*, in: Sternberg, Robert & J. Davidson, Janet (Eds.), *Conceptions of giftedness* (2nd ed.), Cambridge University Press, New York 2005, pp. 98–119.
[31] Bloom, Benjamin, *Developing talent in young people*, Ballantine Books, New York 1985.
[32] Potts, John, *A History of Charisma*, Palgrave MacMillan, London 2009.

Weber's conception of so-called charismatic authority became very famous, so famous that it was imposed retroactively on earlier sources and interpretations. In contemporary sentiment, on the other hand, charisma essentially indicates some mysterious and ineffable quality, probably innate, that characterises certain individuals and makes them attractive in a broad sense; it is, however, attributed to a broader range of personalities than that considered by Weber—who limited himself to considering only political and religious leaders—and includes personalities from show business, sport, and so on. In accordance with Weber's conception, the contemporary one preserves the mysterious quality of charisma. Indeed, for Weber, the irrational nature of charismatic authority represented a breach in the rational view of the world and society constructed by Western modernity. This nature also represented a connection with the original, Christian version of the concept of charisma and represented a reaction to the 'disenchantment of the world'—the loss of ground by any mythological and religious vision—theorised by Weber.

The question, however, is legitimate: does charisma really exist? Or is it a reverie of Weber's, having dreamt up a 'romantic' alternative to the coldness and greyness of bureaucratic and institutional authority? Is it, in short, a remnant of a mystical mentality from the past? There are those who have advanced precisely this critique, such as the French sociologist Pierre Bourdieu, who not only considered this concept naïve, but also regarded it as a theoretical construct designed to justify and shore up power relations. And, in the management sphere, the theorist John Kottler railed against charisma, considering it unessential to effective leadership. In the field of psychology, it was the scholar Len Oakes who carried out a study on eleven allegedly charismatic leaders, finding no particular traits except a certain tendency towards narcissism.[33] Studies subsequent to Weber's have focused, for example, on the followers of the charismatic leader, or have tried to analyse the political effects of charismatic leadership. Regardless of the critics outlined above, in contemporary culture, charisma as a natural quality—and thus as opposed to its artificial surrogate, celebrity—is generally accorded importance. It is no coincidence that, in the field of management literature, texts dedicated to charisma and how to acquire it abound. In short, if our culture continues to maintain this term, to use it, it is obvious that it responds to some expressive need. That is, the need to indicate an unknown factor that cannot be indicated by similar words, such as celebrity, aura, prestige, or whatever.

Let us therefore take a closer look at its historical origins. Few, in fact, of those who use this term today—like, indeed, those who use the term

---

[33] See Oakes, Len, *Prophetic Charisma: The Psychology of Revolutionary Religious Personalities*, Syracuse University Press, Syracuse 1997.

'genius'—are aware of the fact that it goes back to the early Christian tradition, that it is a Greek word that appeared in that context around the middle of the first century AD. With some differences, of course, since the qualities implied in that term indicated miraculous powers of the spirit, such as the ability to speak unknown languages, to make prophecies, to heal illnesses, and so on.

Written around 50–62 AD, Paul's epistles are the first texts to contain the term 'charisma,' used to refer to a gift bestowed by God's grace; the recipients of these epistles then were the very young Christian communities scattered in various urban centres of the ancient world, such as Rome, Corinth and Colossus. Written in Greek, in coining the term 'charisma,' the Pauline epistles take up the Greek root 'charis,' meaning 'favour' or 'grace.' Latinised, the term 'charisma' continued to be present for a while in Christian thought, only to be replaced by other Latin terms, namely, 'gratia' (grace) and 'donum' (gift). In the nineteenth century, German theology regained a certain cultural interest in the early periods of Christianity, and revived the use of the term 'charisma'; this served as an inspiration for Max Weber.

While Pauline charisma is a religious notion linked to a collectivist vision of the first religious communities, Weberian charisma is instead secular and linked to specific individuals, who rise above their communities of origin. But there are also commonalities, most notably the mystical aura that both definitions hold, explicit in the former, implicit in the latter. A form of mysticism, that of the Weberian definition, is intertwined with the most typical aspects of modern society, namely, technology and rationality.

The Greek root 'charis' encompasses multiple meanings: charm, gift, favour, gratitude, benevolence; more specifically, the root in question has three potential meanings around the concept of grace: grace understood as personal beauty, as love or favour, and finally, as a benefit offered not as a reward but out of goodness. This etymological richness derives, in turn, from the different uses of this word over the centuries and, in particular, from ancient Greek epics, classical Greek philosophy, and Hellenistic and Greco-Roman culture. As far as the first source is concerned, it is Homer, in the Iliad, who speaks of the goddess Charis (singular), while the Charites (plural) are mentioned, again by Homer and also Hesiod—according to the latter, the goddesses in question are a trio, also known as Graces. In the Odyssey, Athena grants her charis to Odysseus's son Telemachus, whom the goddess is accompanying in search of his father, so much so that the latter arouses the admiration of all onlookers. St. Paul will take up this aspect—charis understood as a divine gift—when he interprets charisma precisely as the fruit of God's will.

In classical Greece, the meaning of charis concerns both physical beauty and speech, and, again, this quality is of divine origin, to be understood as a favour that binds the receiver and the giver (the deity).

As far as the Hebrew culture—out of which St Paul sprang—was concerned, divine grace was indicated by the root 'hnn,' used in the Old Testament. And, when the Bible was translated into Greek, 'hnn' was rendered as 'charis.' However, the Old Testament also tells us how the power of the divine spirit is manifested to specific individuals destined to become prophets. And, if we wish, we can go even further back in time, connecting the spirit of the Bible with the world of shamanism, and with the idea of a force that permeates material reality and whose access would grant the shaman extraordinary abilities. But all of this would take us too far down a tangential road.

Here, then, are Paul's 'three worlds': the Jewish one—with its prophets—the Greco-Roman cultural one—with its belief in spirits and their ability to influence human affairs—and the Roman juridical one. Paul had dual citizenship, in Tarsus and Rome, which allowed him to move freely throughout the Empire.

Therefore, in need of a keyword to use everywhere, Paul adopted and adapted 'charis'—and thus also 'hnn'—using it according to his needs. He innovated, in short. And he used the concept of charisma to denote a spiritual force that bestows various abilities and helps build and strengthen the community of believers. The term 'charisma' is thus used to denote special abilities that would otherwise have been attributed to magic; it was therefore also an attempt to clearly distinguish himself from the various cults that were raging at that same time in the Roman Empire.

And, after centuries of more or less complete oblivion, this is where Max Weber takes up charisma again; in his work *Wirtschaft und Gesellschaft*—published posthumously in 1922, and translated into English in 1947—the scholar presents charisma as one of the three possible forms of authority—alongside the legal and the traditional. Once the way was opened, the concept of charisma, however, escaped the theoretical work in which it was confined and spread into the world of mainstream culture. And so, as early as the 1960s, the notion of charisma was used in the press to refer to the likes of the Kennedys or Martin Luther King, while later still, this connotation was attributed to actors, singers and various celebrities.

The wholly romantic cult of the genius—that is, of the inspired creator—is also an ingredient that Weber will use in his elaboration of the figure of the charismatic leader. The scholar will therefore extend the notion of genius from the world of culture to that of politics. As for the precise definition

Weber gives, it consists of a certain quality of an individual's personality thanks to which they are considered extraordinary and to which exceptional powers or abilities are attributed. This concept slowly makes its way into the world of culture, and becomes common currency in sociology, political science, and even psychology. And so, if, for historian Arthur Schlesinger, charisma would be irrelevant in democratic societies,[34] Freudian psychoanalyst Irvine Schiffer argues that it actually continues to play a seductive—and even somewhat infantilising—role in current political processes.[35] Schiffer, however, modifies the Weberian conception: for the psychologist, in fact, followers would play an active—and not passive, as Weber suggested—role in the creation of the leader.

The sociologists Jack Sanders and Bryan Wilson hold a similar view of things, in which the leader cannot precisely be charismatic per se, but needs followers who regard him or her as such.[36,37] For Sanders in particular, however, charisma is not simply projected by followers onto the leader, as they do not choose a leader at random, but on the basis of the latter's exceptional qualities.

Not everyone agrees with the idea that charisma has to be a social construct, however. For psychologist Len Oakes, charismatic leaders would possess specific psychological traits that would make them different from everyone else.[38] The scholar—who has dealt with charismatic leaders for years—denies that the likes of L. Ron Hubbard, Bhagwan Shree Rajneesh and Sun Myong Moon represent mere constructs of their followers. Psychometric tests carried out by Oakes on eleven charismatic leaders did not reveal any extraordinary psychological traits; the scholar decided to use a qualitative approach, interviewing twenty charismatic leaders in depth and systematically studying the biographies of dozens of other charismatic figures. And, according to Oakes, charismatic leaders are very energetic and characterised by very high self-esteem, which can border on the pathological. They are able to inspire others with their rhetorical skills and are skilled manipulators; they are also self-centred, i.e., they keep their distance from

[34] Schlesinger, Arthur M. Jr., *The Politics of Hope*, Eyre & Spottiswoode, London 1964.

[35] Schiffer, Irvine, *Charisma: A Psychoanalytical Look at Mass Society*, University of Toronto Press, Toronto 1973.

[36] Sanders, Jack T., *Charisma, Converts, Competitors: Societal and Sociological Factors in the Success of Early Christianity*, SCM Press, London 2000.

[37] Wilson, Bryan R., *The noble savages: the primitive origins of charisma and its contemporary survival*, University of California Press Berkeley 1975.

[38] See Oakes, Len, *Prophetic Charisma: The Psychology of Revolutionary Religious Personalities*, Syracuse University Press, Syracuse 1997.

others, followers included. In short, the line between charismatic leader and narcissistic personality would be a thin one.

Be that as it may, the contemporary use of the term 'charisma' derives mainly from Weber, not St. Paul. Today, it is understood as a characteristic possessed by exceptional individuals, and not a divine gift spread far and wide in various forms over a community of believers. Today, it basically means a personal charm or magnetism that is difficult to resist. Charisma would thus be composed of a certain 'physical presence' and the ability to convince. Other common definitions include 'magnetism,' 'energy,' 'the ability to attract the masses,' but also 'the ability to arouse a sense of inclusive intimacy,' or to 'attract attention by mere presence, even without intention'; beauty, on the other hand, is categorically excluded, in the sense that, to be charismatic, one does not need to be beautiful.

And, since the 1990s, a large number of self-help books have appeared in bookshops that treat charisma as a quality that can be taught and acquired by anyone—just as has long been the case with genius. One example is *The New Secrets of Charisma*, whose author, Doe Lang, argues that charisma is present in varying degrees in everyone, and can be developed in some way, or 'unleashed' through specific exercises.[39] Among the many books that have come out on the subject, let us also mention Olivia Fox Cabane's *The Charisma Myth*, which defines charisma as "presence, power, and warmth," and explains precisely how to obtain these abilities.[40] Obviously, the topic of charisma was too tempting for the academic world to let it slip through its fingers, and, in fact, since the late 1980s, leadership theory, work and organisational psychology, and the world of business research have been dealing with it systematically. In particular, two scholars from McGill University, Jay Conger and Rabindra Kanungo, were the first—in the field of leadership studies—to devote a textbook to the subject of charisma, *Charismatic Leadership in Organisations,* which has since become a classic.[41] Subsequently, the scientific study of charisma exploded, including as a consequence of various empirical studies that highlighted how leaders judged to be charismatic tend to be much more effective, succeed in motivating employees better and generally produce positive effects on the organisation they are part of. Hence, the concept of charisma moves from the field of sociology to that of experimental psychology, resulting in it being subjected to experiments that Weber

---

[39] Lang, Doe, *The New Secrets of Charisma: How to Discover and Unleash your Hidden Powers*, McGraw-Hill, New York 1999.

[40] Fox Cabane, Olivia, *The Charisma Myth. How Anyone Can Master the Art and Science of Personal Magnetism*, Portfolio Hardcover, New York 2012.

[41] Conger, Jay and Kanungo, Rabindra, *Charismatic Leadership in Organisations*, SAGE Publications, New York 1988.

had certainly not thought of. The aim of Conger and Kanungo's research is, in fact, to replace the common view of charisma with a more rigorous conception based on empirical measurements.

Bernard Bass—an American leadership scholar—also puts forward a similar proposal in his *Transformational Leadership*.[42] In this work, Bass considers charisma as one of the components of what he calls transformational leadership; he also develops a Multifactor Leadership Questionnaire to measure the charismatic qualities of those to whom it is administered. Bass also argues that charismatic leadership must be "socialised" rather than "personalised," i.e., narcissism, aggression and impulsiveness must be removed from it, and replaced with an attitude of service to the community.

In a 1989 study, two Canadian researchers, Jane Howell and Peter Frost, argued that charisma is a specific quality that can be identified and separated from other leadership styles. According to them, charisma therefore exists and is not reducible to other known ingredients that constitute the paraphernalia of the successful leader. According to them, charisma is "a qualitatively different phenomenon."[43] Not only that, but the two put forward a bolder hypothesis, namely, that it can be studied in the laboratory, and therefore need not be left in the vague. Howell and Frost also carried out an experiment, based on actors playing three different leadership styles, namely, "structuring," "considered" and "charismatic," under the guise of which they were to ask a group of university students to perform a certain task. The study revealed that the people who worked under charismatic leadership performed their task much better than those led by the other two types of leader. Given, then, that some of the actors were asked to act in a charismatic manner, Howell and Frost conclude, one might think that charisma can indeed be taught, that one can be trained in it.

Another interesting study was conducted in 1988 by Nancy Roberts and Raymond Bradley, in which the authors emphasise a couple of central elements in Weber's conception of charisma, namely, the necessity of a social crisis in order for the charismatic leader to present themselves as a response to it, and the reduction of the role and power of charisma when the change induced by the leader is bureaucratised and framed within a regulated structure. According to Roberts and Bradley, it is a contextual phenomenon and, as such, cannot be intentionally recreated by individuals or organisations, as claimed by other academics, as well as all the self-help books cited. Not

---

[42] Bass, Bernard M. & Riggio, Ronald E., *Transformational Leadership*, Psychology Press, East Sussex 2006.

[43] Howell, Jane M. & Peter J. Frost, Peter J., *A laboratory study of charismatic leadership*, Organizational Behavior and Human Decision Processes, Volume 43, Issue 2, April 1989, pp. 243–269.

only that, charisma could not even be simulated in the laboratory, as claimed by Howell and Frost. In essence, Roberts and Bradley end up supporting the Weberian idea of charisma as an irrational force, which can have unpredictable and sometimes very negative effects, and which it would therefore be better to avoid unleashing.[44] More recently, some leadership scholars have sought to clarify the mystery of charisma without denying its enigmatic nature. In this regard, US psychologist Ronald Riggio has argued that charismatic leaders are characterised by a communicative excellence that combines expressiveness and high empathy. To assess charisma, Riggio developed a measurement system, the Social Skills Inventory questionnaire, which includes six facets related to the charismatic subject, namely, emotionally expressive, enthusiastic, eloquent, visionary, self-confident, and responsive to others.[45] Riggio, however, is still trying to maintain the mystery: according to him, it is by no means certain that possessing these six elements actually makes someone a charismatic person. According to the American scholar Frank Bernieri, charisma is connected to a form of synchronicity between the speaker and the audience, and it is triggered—unconsciously—by small signals. It would be similar to jazz improvisation, i.e., it would depend on acquired skills, but also on the innate ability to adapt to any audience. So, even for Bernieri, charisma cannot really be taught; charismatic people do what they do 'by nature,' but it would still be possible to improve one's communication skills so as to at least come close to this quality. However, it is always important to bear in mind that charisma cannot be attributed to oneself: it is our followers who must attribute it to us. Contrastingly, a lack of charisma immediately leaps to the eye, with the press being quick to point it out: think, for example, of the charisma exercised, for better or worse, by a character like Fidel Castro, and the lack thereof in his brother Raul.

---

[44] Roberts, Nancy C. and Bradley, Raymond Trevor, *Limits of Charisma* in Jeffrey A. Sonnenfeld (ed.) *Concepts of Leadership,* Aldershot, Dartmouth 1995.

[45] https://www.mindgarden.com/144-social-skills-inventory.

3

Towards a Darwinian Theory of Genius

3.1   Evolution and Extreme Creativity

Let us therefore return to the subject of genius, and try to frame this theme within a perspective that, in the course of our research for this book, has undoubtedly become our favourite, namely, the Darwinian one. And, in particular, we introduce the work of Dean Keith Simonton. A professor at the University of California (at Davis), Simonton has devoted his academic career to the study of genius, creativity and greatness in general. The scholar's main work on Darwin and genius is contained in the excellent *Origins of Genius*.[1] In particular, he tries to get to the root of genius, to its foundations, and, far from considering the concept as a mere social construct of the Romantic era destined to fall into oblivion, he frames it in a Darwinian perspective—albeit nothing to do with the alleged genetic or biological nature of genius, mind you. In this chapter, we will try to frame Simonton's Darwinian approach and see if it can in any way help us to unlock the secrets of genius, that is, to understand extreme creativity without mythologising it, but also without trivialising it. This is an important question. It is true that we ordinary people are fascinated by the performances of great geniuses and perhaps cultivate the secret desire to be like them. The study of genius, however, has a broader role, relating to the very nature of our civilisation, and of others. Indeed, human civilisations are often defined in their very nature by the geniuses

---

[1] Simonton, Dean K., *Origins of Genius—Darwinian Perspectives on Creativity*, Oxford University Press, Oxford 1999.

© The Author(s), under exclusive license to Springer Nature
Switzerland AG 2023
R. Manzocco, *Genius*, Springer Praxis Books,
https://doi.org/10.1007/978-3-031-27092-5_3

that shaped them. So important is their contribution that the presence of a certain number of geniuses in a certain civilisation is an indicator of its state of health: a dying civilisation does not, in fact, produce much in terms of creativity and innovation. This is argued by, among others, the eminent American anthropologist Alfred Kroeber, who, in fact, uses the appearance and frequency of creative people in a certain society to assess whether it is growing, stagnating or declining.[2] Let us therefore return to Simonton's work, in the hope that it may help us shed light on the nature of genius. The scholar's approach starts, as is obvious, from certain definitions. In particular, from that of 'creative genius,' which is precisely broken down into two terms: 'genius' and 'creativity.' Let us begin with the first term, leaving aside its historiography—already addressed in the first chapter—and focusing instead on the role of this term within psychology.

Let us therefore start with a lexical definition. Samuel Johnson, writer and author, in 1775, of the first dictionary of the English language, defined genius in his tome—which cost him nine years of work—as "a mind of large general powers, accidentally determined to some particular direction." Starting from this 'dictionary definition,' psychologists have subsequently attempted to identify the actual content of this term, so as to develop an objective system for measuring it, often reducing it to terms that run in parallel, such as 'intelligence' and 'eminence.' We have already seen Binet and Terman's attempts to see genius in terms of intelligence and IQ, and Galton's in terms of eminence. We have also seen how such definitions are insufficient and, however much they overlap with the concept of genius, do not identify with them—otherwise, the greatest genius of humankind would have to be Marilyn vos Savant, an American journalist who entered the Guinness Book of World Records for her IQ of no less than 228 points, the highest ever recorded, and became famous specifically for that score.

And of the eminence dear to Galton, what do we do with it? It certainly helps us, since it eliminates the problem of so-called 'misunderstood' or unrecognised genius. If, in fact, eminence—and therefore reputation—and genius coincide, if a person is not eminent, they cannot be called a genius. Even here, however, there is no shortage of problems: reputation can be misplaced, which is why Galton recommends that it should be reviewed by posterity. There are, in fact, cases—actually not many, according to Simonton's research—in which the judgment of posterity does *not* agree with that of contemporaries: think, for instance, of the case of Frederick Banting and John Macleod, who were awarded the Nobel Prize in 1923 for having

---

[2] Kroeber, Alfred L., *Configurations of Culture Growth*, University of California Press, Berkeley 1944.

isolated insulin. It was later established that credit was due to Banting and his assistant Charles Best, while McLeod's role was secondary. Despite these exceptions, Simonton tells us, contemporaries and posterity almost always tend to agree—for details and statistical background, we refer you to the theoretical work of the Californian psychologist, and, in particular, *Origins of Genius*. Beware, however: to say that genius is identified with eminence is not to reduce genius to the social context, thus relativising it. Eminence, as we shall see in a moment, is not a by-product of the ephemeral fashions of a certain context, but the engine that shapes it. That is to say, for Simonton, there would be objective, or at least less subjective and relativising, systems for evaluating eminence: it would, in fact, tend to remain stable with the passing of generations and eras, a fact that shows how it would not be the fruit of a temporally localised whim. For now, let us limit ourselves to defining this concept as the label attached to all those who have had an impact on history that is widely recognised as lasting and extensive.

The use of the concept of eminence to define, or rather to identify, who is a genius and who is not is very useful: it allows us to include in the definition in question another central characteristic, namely, uniqueness. Genius is distinguished precisely because a person has done something unique, to the point that we can say that every genius is such in their own way. Not only that, eminence is not the same for everyone: there are those who are more eminent and those who are less so. And this allows us to introduce another concept, that of the gradualness of genius: in practice, among geniuses, there are those who are more inclined towards genius and those who are less so. Of course, there are no perfect definitions, and eminence is sometimes subject to the whims of fashion, but our preparatory work allows us, if nothing else, to select the characters—the geniuses—on which to work.

3.2   Originality and Usefulness

As we have seen, the concept of genius is linked to that of creativity: it may be obvious, but we might as well reiterate that, to be classified as a genius, you have to be creative. This consists of two parts: first, what you do—or discover—must be original, at least for the cultural context in which you operate. And so, Galileo's discovery of sunspots is certainly original in the European context, even though China had known about them for thousands of years—a hint of the idea can already be found in the *I-Ching*, the famous *Book of Changes* that dates back to the twelfth century BC, while the first official record dates back to 28 BC. Secondly, the product of creativity must be adaptive. Which means that it must serve a purpose, even if only for aesthetic pleasure. The criterion of usefulness obviously depends on the field we are talking about: an invention—for example, a new propulsion system—must work; a work of art must arouse some emotion; and so on. As far as scientific theories are concerned, they must be consistent and in line with observations; nevertheless, mere originality is not enough.

Hence: the label of creativity does depend on the judgment of the socio-cultural context, but according to measurable criteria, namely, originality—the actual, objective novelty of a certain product for a certain context—and adaptability—the ability to bring about some visibly useful effect. In practice, geniuses are people who have left to posterity a large set of intellectual or aesthetic products that are original and adaptive.

3.3   Enter Darwin

And after these starting definitions, the thread of Simonton's discourse finally reaches the centrepiece, a giant of thought: Charles Darwin. Darwin: a genius who—perhaps—provides us with the necessary interpretative keys to explain it, the concept of genius. His and everyone else's: from Archimedes to Galileo, from Newton to Einstein, and so on. But was Darwin really a genius? Certainly not, if we limit ourselves to defining genius as a very high, "rapid intelligence." In his autobiography, the scientist reminds us, for example, that he was considered by his parents and teachers as a rather ordinary boy, slower to learn than his younger sister, below normal standards of intelligence. He says of himself that he is slow in understanding and lacks sharpness, as well as critical faculties; in this regard, he confesses that, while reading a book or an article, he finds it hard to recognise its weaknesses. He also admits that he is unable to entertain lengthy abstract reasoning, and is therefore unsuitable for mathematics and metaphysics. And, despite all this, there is no doubt that Darwin must be recognised as a genius in terms of the criteria adopted by Galton. Indeed, it is difficult to overestimate the extent of his cultural influence, which literally led the West to revise its view of Man, his origin, his status. Not everyone is given the privilege of turning their surname into an adjective, such as "Darwinian."

Here, then, is the long and tortuous path of Darwin's thought, which starts with original Darwinism, merges with nascent genetics (thus producing the neo-Darwinian synthesis), and then, in an act of cultural imperialism, comes to encompass the social sciences, through Edward O. Wilson's socio-biology. Finally, it ends up inside the brain, with Gerald Edelman's neural

Darwinism—in which the intricate complexity of the nervous system cannot be contained in the genetic code, but must be explained through the Darwinian interaction between neurons and the internal and external environment. So, if Darwinism explains the development of the brain, is it so strange to think—as Simonton does—that genius can also be framed within the paradigm that the English scientist launched?

Let us therefore explore Simonton's proposal in more detail. According to the scholar, the brain is a kind of "Darwinian machine," structured to produce ideas through a process of random variation. Simonton's approach gathers stimuli from various sources, including, at least in part, the work of Burrhus Frederic Skinner. An eminent American psychologist, Skinner was part of the current known as 'behaviourism,' which—at least in his case—even denied the existence of internal mental states. No free will, according to him; no consciousness, no beliefs, nothing. Only observable behaviour. The mind for Skinner was a black box into which external stimuli were introduced and that emitted what he called 'operants,' behaviours that were then selected by the external environment, through positive and negative reinforcement. Put simply, do something useful for the organism and that behaviour will bring you a reward consistent with survival; do something useless or counterproductive and the environment will negatively reinforce that operant, making it less likely in the future. In practice, this theory—which has many Darwinian aspects—does a good job of explaining how articulate behaviour is formed in humans and animals without reference to any genetic programming. Even creative behaviour would be the result of this procedure, which Skinner calls "operant conditioning." Simonton's work certainly does not stop with Skinner, but circumvents the behaviourist concept of the black box and, by referring directly to Darwinism, calls into question the ideas of, among others, Donald T. Campbell—the father of so-called 'evolutionary epistemology'—and the evolutionary biologist Richard Dawkins—who developed a concept that everyone is now familiar with, thanks to the Internet, namely, that of the 'meme.'[3] But let us go in order.

For Simonton, we have first-level Darwinism, which is to be understood in a literal sense, and which concerns the world of life. This includes all of the theories and research concerning evolution, living forms, organs—including the brain—and so on. Then we have a second-level Darwinism, to be understood as a paradigmatic metaphor, i.e., as a tool to be used to interpret objects other than classical living forms. For example, the various human cultures, which can precisely be interpreted as entities that evolve and have aspects

---

[3] The term 'meme' is an abbreviation of the classical Greek word 'mímēma', which means "imitated thing."

adaptive to the environments in which they arise. And so, adaptations developed by living organisms through mutation and selection are then acquired by a certain species through genes; cultural variations, on the other hand, become part of a certain culture in the form of widely disseminated ideas, which are propagated from one mind to another through education and popular traditions and which we can define precisely as 'memes.' According to Dawkins, memes include everything that makes up culture: ideas, songs, fashions, technologies, myths, fables, works of art, philosophical thoughts and whatnot. Of course, this does not mean that Darwinism can be slavishly applied to culture: memes and genes are not the same thing, not least because the former are transmitted between individuals and not just from one generation to the next. Not only that, they can even go in the opposite direction than genes, such as when we observe teenagers teaching their parents how to use computers and mobile phones. Cultural evolution also tends to be very fast, unlike organic evolution, which takes a very long time. Nevertheless, we will pursue this comparison between organic evolution and cultural evolution, pointing out that, while, in the former case, it is the genes that undergo the well-known process of blind variation and selection, in the latter, it is the memes. And, far from resting on the anti-mentalist assumptions of behaviourism, Simonton underlines the fact that, as far as memes are concerned, the aforementioned process of variation and selection takes place in a specific place: within us. For the scholar, the human mind does not deal with new situations merely by producing innovative behaviour, but seeks to anticipate them by producing cognitive representations of possible situations and responses, then testing them against the internal representation of the world that we all develop over time. So, let us be careful not to overdo our reliance on Darwinism, but to use it whenever it is possible to explain a human phenomenon with elegance and simplicity. As is precisely the case with genius. If we define creativity as the production of ideas that are both original and adaptive, the act of creation can then be grafted onto the Darwinian paradigm outlined above and be defined as a special case of the process of random variation and selection. The creative person is the one who generates a lot of new ideas and then subjects them to selection according to technical, intellectual or artistic criteria. In other words, to create is to produce memes, and the genius, according to Simonton, is the one who produces a vast assortment of memes to be bequeathed to future generations. If, therefore, Darwinianically 'successful' organisms propagate their genes in the form of numerous offspring, Darwinianically 'successful' geniuses are individuals who propagate numerous memes. Reproductive success versus productive success. An idea that, by the way, again elegantly resolves the issue of forgotten geniuses: if one is truly a genius, one's memes will eventually propagate sooner or later, despite possible difficulties, and thus the concept of a 'forgotten genius' would be a contradiction in terms: at best, one may be temporarily forgotten, but will, sooner or later, be rediscovered.

Thus, Simontonian thought is perfectly rooted in the tradition of so-called 'evolutionary epistemology,' a system of thought that seeks to construct a 'Darwinian theory of knowledge,' that is, a system of thought that philosophically interprets the human act of knowing in evolutionary terms. In two ways: firstly, by offering an organic, biological explanation of the human mind, of so-called 'common sense,' and also of the limits of our cognitive apparatus; secondly, in metaphorical and analogical terms, i.e., as a general model within which to frame the social sciences and our cultural history. It is also true that everything, in a certain sense, can be seen in epistemological terms, i.e., in terms of the theory of knowledge. For example, as one of the founders of this approach, Konrad Lorenz—the father of ethology, the science of comparative human and animal behaviour—argued, in a certain sense, the wings of birds embody knowledge relating to the laws of aerodynamics, in the sense that their structure reflects the external world, and thus "knows it." And one of the leading exponents of this approach was Donald T. Campbell, according to whom the creative process itself can be read in Darwinian terms. In this case, creativity would concern three aspects. First, there would be a mechanism capable of randomly generating ideas of many kinds, and in large quantities—a bit like the mutations that occur in the natural world. These 'mutations of the mind' would then be subject to a selection process—that is, there would be a further selective mechanism, analogous to the selection that the environment uses in regard to organisms, that would be cognitive and cultural in nature. Finally, there would be a process of retention of the selected ideas, which would then accumulate in the individual memory, and also in the collective memory. In essence, the act of creation would be a trial-and-error process, fuelled by cognitive rumination—'thoughts in the wild'—and behavioural experimentation.

3.4 Darwinian Anecdotes

Simonton, at this point, goes in search of authors who have described their own creative process in ways that actually resemble the procedure of random variation and selection. And he finds several. French poet, philosopher and writer Paul Valéry states that "it takes two to invent anything. The one makes up combinations; the other chooses, recognises what he wishes and what is important to him in the mass of the things which the former has imparted to him."[4] For his part, the seventeenth-century English poet and playwright John Dryden described the composition of a work poetically, saying that it began "when it was only a confused mass of thoughts, tumbling over one another in the dark; when the fancy was yet in its first work, moving the sleeping images of things towards the light, there to be distinguished, and then either chosen or rejected by the judgment."[5] And do not think that this kind of description only concerns literary types and artists. The famous nineteenth century English scientist Michael Faraday, for example, tells us that "the world little knows how any thoughts and theories which have passed through the mind of a scientific investigator have been crushed into silence and secrecy by his own severe criticism and adverse examinations."[6] Even Nobel Prize winner Linus Pauling used to say that "the way to get good ideas is to get lots of ideas and throw the bad ones away." So, while organic evolution produces variation by recombining genes, creativity is about recombining ideas, possibly in unexpected ways. This aspect— Simonton tells us—was already identified by an Austrian scientist about a hundred years ago: we refer to Ernst Mach, who described himself as having "a powerfully developed mechanical memory, which recalls vividly and faithfully old situations," but that "more is required for the development of inventions (...) more extensive chains of images are necessary here, the excitation by mutual contact of widely different trains of ideas, a more powerful, more manifold, and richer connection of the contents of memory."[7] In short, it seems that, for Mach, invention and imagination proceed according to a large degree of randomness, just as Campbell and Simonton's theories demand. And the French mathematician and philosopher Henri Poincaré also confirms this process. Recalling an evening of insomnia, due to a late cup of coffee, the scholar recounts that "ideas rose in crowds; I felt them collide until pairs interlocked, so to speak, making a stable combination. By the next morning I had established the existence of a class of Fuchsian functions." Interesting, then, is Poincaré's comparison of these colliding images with "the hooked atoms of Epicurus," which collide "like the molecules of gas in the kinematic theory of gases" to the point that "their mutual impacts

---

[4] Quoted by Hadamard, Jacques, *The Psychology of Invention in the Mathematical Field*, Princeton University Press, Princeton 1945, p. 30.

[5] Dryden, John, *Epistle dedicatory of The Rival Ladies*, in: W. P. Ker (Ed.), *Essays of John Dryden* (Vol. i, pp. 1–9), Oxford: Clarendon Press Oxford 1926 (original essay published 1664).

[6] Faraday, Michael, *Experimental Researches in Chemistry and Physics,* p.486, 1859.

[7] Mach, Ernst, *On the part played by accident in invention and discovery,* Monist, 6,161–75, 1896.

may produce new combinations."[8] Added to this is the deeply visual and vivid nature of this process. For example, according to Nobel Prize winner Max Planck, a scientist "must have a vivid intuitive imagination, for new ideas are not generated by deduction, but by an artistically creative imagination."[9] Albert Einstein, for whom imagination, not words, is what counts, is on his side: "The words of the language, as they are written or spoken, do not seem to play any role in my mechanism of thought. The psychical entities which seem to serve as elements in thought are certain signs and more or less clear images which can be 'voluntarily' reproduced and combined (…) From a psychological viewpoint this combinatory play seems to be the essential feature in productive thought (…) The (…) elements are, in my case, of visual and some of muscular type. Conventional words or other signs have to be sought for laboriously only in a secondary stage, when the mentioned associative play is sufficiently established and can be reproduced at will."[10] This is a position reinforced by French mathematician Jacques Hadamard, for whom "as to words, they remain absolutely absent from my mind until I come to the moment of communicating the results in written or oral form."[11] Which goes a long way toward explaining the difficulties that great thinkers—scientists, philosophers and whatnot—encounter in putting their insights into words.

3.5   Genius and Dream States

To this, we would also add that, from what has emerged so far, creating and dreaming seem to be two somewhat connected processes, that the images that populate the minds of great creative geniuses have some dreamlike aspect. And psychiatry seems to agree: the psychiatrist Albert Rothenberg, in particular, has studied the aforementioned phenomenon, articulating it—according to him—in two processes, "homospatial thinking" and "janusian thinking."[12] The first process consists in imagining two objects or entities as occupying the same space, leading to the formation of a new entity/identity—in essence, we are dealing here with the cognitive fusion and visual superimposition of different things, leading to the birth of new discrete entities. The

---

[8] Poincaré, Henri, *The foundations of science: Science and hypothesis, the value of science, science and method*, Science Press, New York 1921, pp. 387–393.

[9] Planck, Max, *Scientific autobiography and other papers*, Philosophical Library, New York 1949, p. 109.

[10] Einstein, Albert, *Ideas and Opinions*, Crown Publishers, Inc., New York 1960, pp. 25–26.

[11] Hadamard, Jacques, *The psychology of invention in the mathematical field*, Princeton University Press, Princeton 1945, p. 82.

[12] Rothenberg, Albert, *The emerging goddess: The creative process in art, science, and other fields*, University of Chicago Press, Chicago 1979.

second process, on the other hand, consists in the simultaneous conception of two opposing ideas—the term 'Janusian' derives from the Roman god Janus, famously endowed with two faces looking in opposite directions. And, according to Rothenberg, it is thanks to Janusian thought that Niels Bohr arrived at the principle of complementarity and the idea that light can be both wave and particle at the same time—a very Janusian idea, indeed. And, on this subject, the Danish physicist—in his essay *Discussion with Einstein on Epistemological Problems in Atomic Physics*—relates that "In the Institute in Copenhagen, where through those years a number of young physicists from various countries came together for discussions, we used, when in trouble, often to comfort ourselves with jokes, among them the old saying of the two kinds of truth. To the one kind belong statements so simple and clear that the opposite assertion obviously could not be defended. The other kind, the so-called 'deep truths,' are statements in which the opposite also contains deep truth."[13] At this point, Simonton makes a recommendation, and advises us not to overdo this insistence on the imaginative aspects of creative thinking: in particular, the scholar cites a study carried out on 64 leading scientists, according to which the majority of their mental processes would consist of "imageless thought," in particular, before an important discovery or intuition.[14] It would therefore be a process that, at least in its final stages—characterised by a period of incubation—takes on an unconscious connotation, while the previous conscious work of preparation would be aimed at setting things in motion within the mind—untangling the 'atoms' of the mind and shuffling them, Poincaré might say.

Until the moment of illumination: the very famous "Eureka," Archimedes' "I found it!". And also that of Darwin, who, with regard to his evolutionary intuition, tells us in his autobiography that "I can remember the very spot in the road, whilst in my carriage, when to my joy the solution occurred to me."[15] For his part, the father of Egyptology, Jean-François Champollion, when, on September 14, 1822—around midday—he finally managed to decipher the Rosetta Stone, rushed into his elder brother's office and threw some depictions of Egyptian inscriptions on the latter's desk to the cry of: "Je tiens l'affaire!" He then fainted, waking up five days later.[16]

---

[13] Schilpp, Paul Arthur (Ed.), *Albert Einstein: Philosopher-Scientist*, The Library of Living Philosophers, Vol. VII, Evanston, Illinois 1949, pp. 201–241.

[14] Quoted in: Roe, Anne, *The making of a scientist*, Dodd, Mead, New York 1952, p. 144.

[15] Darwin, Charles; Darwin, Francis, *Autobiography and Selected Letters*, Courier Corporation, North Chelmsford, Massachusetts 1958, p. 43.

[16] This is told by his biographer, German Egyptologist Hermine Hartleben. Quoted in. Robinson, Andrew, *Sudden Genius? The Gradual Path to Creative Breakthroughs*, Oxford University Press, Oxford 2010, p. 124.

3.6   Insight and Serendipity

Now, there is one type of discovery/intuition that particularly attracts Simonton's interest, and that is what is known as 'serendipity,' i.e., the lucky discovery, made by chance, while one is busily searching for something else unrelated. Famous in this respect is, at least among the general public, the case of Viagra, discovered while the team working on it was trying to come up with a drug to treat high blood pressure. The term 'serendipity' comes to us from the eighteenth-century English writer Horace Walpole, who refers us to an ancient Persian fable about *The Three Princes of Serendip*—whose title characters go on various adventures, discovering unexpected and unwanted things, by chance and through mental acuity. An interesting concept, that of 'serendipity,' one that was later taken up by the scholar Walter Cannon—who invented the concept of the 'fight or flight response'—in his essay entitled *The Role of Chance in Discovery*[17] Beyond the magic blue pill, the history of science is not lacking for cases of serendipity. For example, there is the well-known case of the father of modern printing techniques, Johannes Gutenberg, who had the intuition of the printing press while observing a wine press during a grape harvest. Basically, a lucky chance that gave him the right idea. Or think of Christopher Columbus, who certainly did not aim to discover a new continent, but rather to find a new route to the Indies. Simonton then reminds us that, sometimes, it is not even necessary to have a precise idea in mind; this was the case, for example, for the Dutch scientist Antonie van Leeuwenhoek, the father of modern microbiology, who, out of sheer curiosity, observed various substances under a microscope and discovered numerous microscopic organisms. Serendipity is so important in the world of scientific discovery that Simonton suggests an interesting notion, namely, that it depends on the encounter between chance and the particular openness to external stimuli of creative geniuses. That is, the creative genius is as such insofar as it notices something that another, less prepared mind does not take into consideration.

Our excursus into the world of Darwinian creativity finally brings us to a concept that many identify with the creative act itself, namely, the so-called 'insight.' This phenomenon, which is identified with the notion of sudden enlightenment, has stimulated a great deal of psychological research, particularly in the field of Gestalt Psychology—a current that originated in Europe at the beginning of the twentieth century and focuses, in particular, on perception, although also on mental processes. And, since we are talking

---

[17] Cannon, Walter, *The role of chance in discovery*, The Scientific Monthly 50.3 (1940): 204–209.

about Darwin and organic evolution in this chapter, let us mention the fact that, according to Gestaltists, animals are also subject to insight. Famous in this regard are the experiments with chimpanzees organised by the gestalt psychologist Wolfgang Kohler—described in his classic work *The Mentality of the Apes*[18]—in which he shows that these animals also undergo a sudden and original 'reorganisation of experience,' i.e., insight. In this case, the experiments in question concern chimpanzees faced with a problem—how to acquire a banana that is out of their reach—which is solved after a period of incubation through a reorganisation of perceptual experience—a few crates placed one on top of the other in order to reach the coveted goal. If, for gestaltists, this process allowed the animal—or human—to circumvent the trial-and-error process, according to Simonton, this would not be the case, precisely because, according to him, chimpanzees' minds would precede this insight with numerous ideational variations. In fact, the right solution would generally be preceded by wrong solutions, in any case. And, as a human example of this process—an insight preceded by unsuccessful attempts and an incubation period—Simonton cites the case of Carl Friedrich Gauss; the mathematician, after trying unsuccessfully, literally for years, to solve a certain problem, tells us that "finally, two days ago, I succeeded, not on account of my painful efforts, but by the grace of God. Like a sudden flash of lightning, the riddle happened to be solved. I myself cannot say what was the conducting thread which connected what I previously knew with what made my success possible."[19] What triggered the insight may have been, according to Simonton, some arbitrary event that pointed Gauss's unconscious in the right direction, after a long period of unconscious incubation in which the super-focused attention of the conscious mind relaxed, perhaps allowing for the perception of some subtle semi-conscious indication from his surroundings, some external stimulus that triggered the insight.

Not only that, but the best creativity—Simonton swears—would be serendipitous, not deliberate. And the psychologist cites, in this regard, the German surrealist painter Max Ernst, a proponent of the artistic technique of frottage—a procedure that consists of the 'automatic,' i.e., spontaneous, rubbing of a pencil on an uneven sheet of paper in order to achieve unexpected results. Ernst tells us at one point that: "I was struck by the obsession that showed to my excited gaze the floor-boards upon which a thousand scrubbings had deepened the grooves … [I]n order to aid my meditative and hallucinatory faculties, I made from the boards a series of drawings by placing

---

[18] Kohler, Wolfgang, *The Mentality of Apes,* London: Kegan Paul, Trench, Trubner. U.S. Edition 1925 by Harcourt, Brace & World.

[19] Quoted in: Simonton, Dean K., Op. Cit., p. 44.

on them, at random, sheets of paper which I undertook to rub with black lead. In gazing attentively at the drawings thus obtained... I was surprised by the sudden intensification of my visionary capacities and by the hallucinatory succession of contradictory images superimposed, one upon the other."[20] What we have detailed so far might make you think that creativity falls upon human beings utterly randomly; in reality, we must not forget that our species is characterised by volition, that is, by reasoning in terms of goals, and that it is precisely this way of being that sets the creativity machine in motion. In short, random variations yes, but aided by individual motivational predisposition. Or, in the words of Louis Pasteur, "chance favours only the prepared mind."

And what happens after we have a good intuition? Do we rest on our laurels? Certainly not. The creative genius is forced to work hard on the product of their mind, refining it, demonstrating it, testing it, and, above all, making it presentable to the—often very demanding—community of their peers. Hard work, in short; blood, sweat and tears. So: brilliant creativity represents a mixture of sudden revelations and craftsmanship aimed at refining initial ideas. Of course, from the point of view of spectacularity, we tend to remember sudden insights more; but craftsmanship plays an equally important role, and the career of a genius must consist of both elements. Indeed, ideas arrived at step by step are often more common than those manifested suddenly.

3.7 The Asymmetry of Creativity

Let us now return to the core of Simonton's Darwinian theory of creativity, and ask ourselves: why did we develop the capacity to be creative, indeed, brilliant? The most obvious answer is that creativity plays some adaptive role. Faced with a problem, we try the known solutions, but, if these do not work, we try something less usual, progressively moving away from the known and venturing into territories—i.e., solutions—that are less and less ordinary.

There is a problem, however, and not a small one. From a practical point of view, we, in fact, have a decidedly asymmetrical distribution of creativity, whereby the majority of the human population exhibits little or no such ability, while a very small percentage exhibits a very large amount. It is not clear how such an ability could therefore play an adaptive role, because—if it did—many more people would exhibit an extreme degree of creativity. We

---

[20] Ibid. p. 46.

have no convincing answers, but we can try to hazard some vague hypotheses. We could, for instance, hypothesise a complex interplay between biological and cultural evolution, such that we could say that exceptional creativity cannot be separated from a broad social support network, so that, for every highly creative individual, there must necessarily be many individuals who are not as creative but who provide support—money, food, and so on—to the former.

And how would these Darwinian hyper-creative individuals be more precisely distinguished from the mass that supports them? To begin with, the hyper-creatives produce—as one might guess—a lot, i.e., they leave behind a large number of contributions. Although, in reality, within the world of the creative, we are dealing with a continuum that ranges from a minimum of one idea—Gregor Mendel, whose fame is only due to his experiments on the genetics of peas—to an unspecified, but nonetheless vast, number of contributions—just think of Darwin, who published almost one hundred and twenty publications on the most disparate topics. Not that the number of contributions is the only factor of eminence; on the contrary, quality also matters, of course. Simonton cites the case of John Edward Gray—who?— an English zoologist of more or less the same era and author of no less than 883 publications—many more than Darwin's 119, then, and a vast universe compared to Mendel's paltry seven.

3.8   Proteus, Machiavelli, Darwin

Simonton puts forward a further hypothesis, namely, that somehow geniuses must possess a 'Darwinian personality,' i.e., common psychological traits that favour the production of a large number of ideas, from which to select those that will survive. A minimum of intelligence is necessary, of course: certainly, no creative genius has a below-average level of intellect. Intelligence counts for relatively little, however, in the view of this hypothetical Darwinian personality; what would count instead would be the ability— underlined as early as the 1960s by the studies of the American psychiatrist Sarnoff Mednick—to make very remote associations between separate ideas. According to the psychiatrist, creative people have what he calls a "flat hierarchy of associations," as opposed to the "steep hierarchy of associations" of non-creative people.[21] Bear in mind that this characteristic has nothing to

---

[21] Mednick, Sarnoff A., *The associative basis of the creative process*, Psychological Review, 69,220–32, 1962.

do with individual intelligence: one can be extremely intelligent and uncreative at the same time, according to Sarnoff and Simonton. That is, two people can have the same cultural background—the same memes, if we want to put it that way—but a very different level of associative richness. This is an idea similar to that proposed by the aforementioned Guilford, who distinguishes precisely between convergent thinking—classical analytical and rational thinking—and divergent thinking—the ability to generate multiple alternative responses.

Apart from that, what other characteristics does the genius personality have? For Simonton, there is no doubt: geniuses must possess an excellent tolerance of ambiguity—the ability to 'live with' events and situations that are not clearly defined.[22] They seek novelty and complexity, are open to diversity and are able to practise a form of attention known as 'defocused attention'—the ability to pay attention to a very large amount of information without filtering it and slowing down its processing. More specifically, defocused attention allows the mind to pay mild attention to multiple ideas, even those that are not connected in any way to each other. Geniuses also have very broad interests that are not confined to their specific field of creative action. They tend to have an introverted personality, to the point of becoming anti-social in some cases. Self-centred and independent, they often challenge norms and conventions, sometimes presenting themselves as rebels. Other characteristics of the Darwinian creator are a commitment seasoned with enthusiasm, dedication, persistence—useful for countering defeats and disappointments—and flexibility—that is, there is in them a strange mixture of conviction and pragmatism, which allows them to handle creative issues that are difficult to solve. In the words of B. F. Skinner: "a first principle not formally recognised by scientific methodologists: when you run onto something interesting, drop everything else and study it."[23] This is a rough description, of course, not a perfect portrait of genius. Above all, the traits in question may vary, so, for example, a certain creative genius may not have the highest intellectual capacity, but compensates with the purest of motivations, and so on.

---

[22] A concept originally developed by the Austrian-Polish scholar Else Frenkel-Brunswik. See Frenkel-Brunswik, Else, *Intolerance of ambiguity as an emotional and perceptual personality variable,* Journal of Personality, 18(1), 108–143, 1949.

[23] Skinner, B. Frederick, *A case study in scientific method,* in: Koch, S. (Ed.), *Psychology: A study of a science* (Vol. 2), McGraw-Hill, New York 1959, pp. 359–379.

As far as introversion is concerned, this would be a necessary ingredient, given the need for the creative genius to work long hours in solitude on the issues he or she deals with. Speaking instead of defiance of norms and conventions, let us emphasise that this concerns both social conventions and those internal to the discipline to which the genius in question contributes. And so, for example, we cite the famous case of Niels Bohr, who, as is well known, in 1958, at a conference at Columbia University in New York, said to Wolfgang Pauli, after the latter had presented the nonlinear field theory of elementary particles that he had developed with Heisenberg: "We are all agreed that your theory is crazy. The question which divides us is whether it is crazy enough to have a chance of being correct. My own feeling is that it is not crazy enough." An attitude that led Bohr to reject the usual criteria for what a well-crafted scientific theory should look like and to promote revolutionary concepts such as the complementarity principle.[24] If you are a lover of Greek mythology, you will no doubt be familiar with the myth of Proteus, an ancient marine deity who represented the constant mutability of the waters and who was, in fact, capable of constantly changing shape. From this trait of Proteus, the adjective 'protean,' which indicates flexibility, was derived, and has also entered psychology to indicate a particularly adaptable and hard-to-figure personality type. If there is one trait that seems to characterise the genius personality, it is precisely that of 'proteanism.' The whole thing stems from another concept, that of 'Machiavellian intelligence.' According to the latter idea, primates—including us, then—due to their social nature, have been forced by evolution to develop very sophisticated cognitive and strategic skills; in practice, they have learned to play politics, so as to outsmart their rivals, and thus gain all kinds of advantages. In particular, they have learned to lie, to conceal their intentions, and to pretend. And, among these tricks, we can count so-called 'proteanism,'[25] that is, the ability to be unpredictable, to not behave according to a specific script, but to choose instead behaviour that we could define—at least in appearance—as random, and thus that is in line with Simonton's theory on the nature of random variation in the creative act.

---

[24] The principle of complementarity holds that subatomic objects have certain pairs of complementary properties—e.g., position and momentum—that cannot be observed—and measured—simultaneously.

[25] See Miller, Geoffrey, *Protean Primates: The Evolution of Adaptive Unpredictability in Competition and Courtship*, in: A. Whiten and R. Byrne (Eds.), *Machiavellian Intelligence II*, Cambridge University Press, Cambridge 1997, pp. 312–340.

3.9   Highs and Lows of Creativity

We enter at this point into one of the most debated and questionable aspects of Simonton's approach. Indeed, according to the American scholar's research, it would appear that creativity, including the production of masterpieces of all kinds, depends on chance. More specifically, quality—the works of genius—would only be a probabilistic function of quantity—the number of works produced. This is a controversial assertion, one that does not fully convince us, perhaps because we are still inexorably attached to a more classical idea whereby genius would have far more control over its works than Simonton is willing to attribute to it. It is also the view of Romanticism, for which the artist was able to create works through his or her willpower—admittedly, a supernatural willpower, inspired by some super-individual element, but nevertheless connected to his or her individual psyche. While the Romantic view is not taken seriously in our modern conception [This was my best interpretation of your meaning.], that of performance psychology—which we will examine more closely in a later chapter—is instead examined with care. In particular, it is well known that cognitive psychology has studied performance in the most diverse fields, from music to sport to chess. It is precisely from this research that notions such as the idea that mastery in a certain field is achieved through the famous ten thousand hours—or ten years—of intense deliberate practice have emerged. The question raised by Simonton is crucial: is it possible to consider creativity as a form of expertise, that is, as a skill that can be learned and, above all, improved through practice? No, according to him. Sure, there are the basics of the discipline one practices to be acquired, but, after that, creativity would only be the probabilistic fruit of mere productivity, otherwise we would observe consistent levels of creativity in the world. In other words, a creative person—a composer, for example, or a visual artist—should constantly produce masterpieces. In reality, there are very few creatives who produce nothing but masterpieces on a continuous cycle: the vast majority of geniuses alternate between works of higher value and works of lower quality. It is difficult to object to this statement. All of this does not mean that there is no learning in the field of creativity: creators can still learn tricks and techniques to meet the minimum standards necessary to make a certain product presentable—a paper suitable for publication, a play suitable for presentation, and so on. Indeed, it is certainly possible that these minimum criteria are maintained by the creative person throughout his or her life. What is important for Simonton, however, is the fact that being a genius does not mean being infallible, that is, that one cannot avoid a life of creative ups and downs.

4

Genius and The Enchanted Loom

4.1   For a Neuroscience of Genius

Let us begin with the usual preamble that one makes when talking about the brain: the brain is still a mysterious organ, we know little about it, but, in recent years or decades, giant strides have been made, including thanks to the development of technologies such as MRI and the like. Or we could also start with the famous metaphor conjured by the English Nobel Prize-winning physician Charles Scott Sherrington, who called the brain an 'enchanted loom,' to indicate its complexity. Having said all that, let us now take a look at what neuroscience has to say about creativity and genius, bearing in mind that it is—lo and behold—a complex phenomenon, which is indeed difficult to deal with, so that relatively few contemporary neuroscientists seem to have dealt with it—even though interest in the subject has been growing lately.

And, among these few, we would certainly mention the interesting work of Nancy C. Andreasen, an American neuroscientist and neuropsychiatrist, who, among other things, has also dedicated a popular book to the subject that we recommend, *The Creating Brain*.[1] What interests us in this chapter is to use Andreasen's research to understand—if possible—what happens inside the brain of a genius when they produce their works. The act of creation, in short, but from the point of view of the nervous system. Obviously, in this case, the traditional definition of creativity, linked to the Christian idea of creation—it is no coincidence that Giorgio Vasari, referring to Michelangelo,

---

[1] Andreasen, Nancy C., *The Creating Brain: The Neuroscience of Genius*, Dana Press, New York 2005.

© The Author(s), under exclusive license to Springer Nature
Switzerland AG 2023
R. Manzocco, *Genius*, Springer Praxis Books,
https://doi.org/10.1007/978-3-031-27092-5_4

called him "divine"—does not fit. We need to go deeper. And the scholar's reflection also starts from the Termites, that is, the first attempt to study genius and creativity, precisely weakened, however, by the fact of having defined these gifts as equivalent to the concept of "high intelligence." Our Terman, himself a precocious child, wanted to test precocity in children—about whom a somewhat inauspicious hypothesis has always hovered, known as 'early ripe, early rotten'—demonstrating in his doctoral thesis that abilities developed very early did not necessarily end up declining rapidly.[2] This was the ideal starting point for the six mighty volumes that originated from his research project.

Already in Terman's study, according to Andreasen, a clear difference can be seen between intelligence—with which the children were endowed, and which later provided them with material and social well-being—and creativity—few children shone for such endowment in the sciences and the arts.

Andreasen also concludes that intelligence and creativity, although connected, are not the same thing. Which is refreshing, since it is also our opinion—a fact that might convince you readers to label the writer a "victim of confirmation bias." More specifically, it would be necessary to reach a certain level of intelligence, meaning that IQ and creativity would run in parallel up to a certain point, after which other factors would come into play. In practice, creativity would be the result of a mechanism semi-independent of intelligence, one that enables some people to create scientifically, technologically, artistically, and so on. And here arises—more or less spontaneously—the question mentioned above: in order for a certain work to be defined as creative, does it need external, social or otherwise intersubjective confirmation? This is quite a problem, since if we deal with neuroscience precisely, we are interested in objective phenomena, which should not need social validation in order to be considered as existing. In other words, before constructing a neuroscience of creativity, we must ask ourselves *when* the creative act occurs: when do we have insight? When do we process it and define its details? And when is our contribution accepted by society and defined as creative or ingenious? Emblematic, from this point of view, is the case of Vincent van Gogh, who painted literally thousands of works in the course of his life, without any of them being bought or considered worthy of admiration—except in a few cases, thanks to his friends. After his death,

---

[2] If nothing else, Terman's initial research showed that, far from being—according to the stereotype accepted at the time—emotionally fragile and unfit for social relationships, the precocious children were actually healthier and stronger than those in the control group. Moreover, they were, on average, psychologically balanced, without any particular problems.

however, his works were revalued, to the point of costing millions of dollars. Now, many will tend to think that the act of creation occurred at the moment van Gogh painted these works; not everyone agrees, however, and there are those who argue that creation should actually be measured by external, independent, and perhaps socially accepted criteria. The greatest proponent of the latter approach is the Hungarian-American psychologist Mihaly Csikszentmihalyi; according to him, creativity can never be considered the prerogative of the individual mind, but must necessarily pass through the gauntlet of evaluation by experts in a certain field, who decide whether or not to include it in the cultural domain to which it belongs.[3] More precisely, Csikszentmihalyi argues that creativity is the result of the dynamic interaction among three elements, namely, the domain, the field and the person. When speaking of interaction, it is inevitable that the concept of the system comes into play. Developed by various authors—including the Austrian biologist and philosopher Ludwig von Betalanffy—this notion consists, in simple terms, of the idea that reality is structured on levels of increasing complexity, and that each level has emergent properties that cannot be reduced to, and cannot be explained by, those of the lower levels. To simplify further: water possesses properties that are not reducible simply to those of hydrogen and oxygen, but are indeed "emergent." Now, for the Hungarian-American scholar, there would be the aforementioned three elements from which creativity would emerge. As far as the role of the individual is concerned, creativity would indeed manifest itself due to the latter's work, but this contribution would only emerge, and thus be 'truly' creative, after systemic interaction with the field—in practice, the set of experts in the field—and its inclusion in its relative domain—the knowledge acquired by the field in question. Certainly an elegant theory, but one that leaves us full of doubts. We have talked about van Gogh. Let us now mention the American poet Emily Dickinson, who published very little during her lifetime—ten poems and a letter—Gregor Mendel, whose work on genetics was only recognised after his death; Copernicus, who kept his astronomical studies secret for fear of ecclesiastical reprisals. Now, according to the systemic model, these geniuses would not have been truly creative in life, a statement that, frankly, leaves us somewhat perplexed. Also puzzling is Andreasen, according to whom what counts is the process that the brain goes through when it creates something, such as a poem, or a scientific theory. She assures us that neuroscience will sooner or later provide us with a comprehensive answer, although it is by no means clear how and when.

---

[3] Csikszentmihalyi, Mihaly, *Creativity: Flow and the Psychology of Discovery and Invention*, Harper, New York 2013.

4.2   Dream and Creation

Let us begin, then, to delve into the neural roots of creativity and genius, but to do so, let us take a step back, and start with dreams, or rather a specific dream, induced by opium consumption. According to the English poet Samuel Taylor Coleridge, one day in 1797, he was reading a passage about Xanadu[4] in *Purchas his Pilgrimage*,[5] he fell asleep, and had a dream—the direct result of opium consumption—that kick-started the writing of the poem known as *Kubla Khan: Or, A Vision in a Dream*. Coleridge's account is one of the most famous descriptions of the creative process available to us, while the poem in question—written at the beginning of his poetic career, at the age of twenty-five, and considered perhaps his finest work—was never completed. In any case, it describes the construction in Xanadu, by the Emperor Kubla Kahn, of a marvellous palace surrounded by walls. The central part of the poem presents us with the act of creation depicted as a violent process beyond human control, with elements at once pleasurable, sacred and frightening. According to the poet's account, his brain, sedated and dreaming, finds itself spontaneously producing, without conscious control, a jumble of images and verses that constitute the poem in its entirety. Unable to transcribe the whole poem, Coleridge will attempt to complete it using conscious efforts, only to end up failing. An event, one of failure, that also prompts neuroscience to ask questions about the role of spontaneity, the unconscious and preliminary preparatory work in the act of creation.

You don't think we're going to point out the exact spot in the brain where the creative act originates, though, do you? We have said that we have to take a broad view, not least because we are dealing with a systemic phenomenon, which requires a top-down approach, in which various disciplines—biography, for example, the accounts of creative people, or psychiatry—are used to get an idea of what goes on inside the brain. Here, too, then, we have to resort to definitions; in particular, we have to ask ourselves whether creativity is—in statistical terms—"discrete" or "continuous," that is, whether it progresses in stages or whether it presents "leaps." This is also important from a neuroscientific point of view, because knowing this would give us an indication as to what goes on inside the brains of creative people. Are the same things

---

[4] Xanadu, located in Inner Mongolia and now known as Shangdu, was the summer capital of the Yuan dynasty before Kublai Kahn decided to move the throne to what would become modern Beijing.

[5] *Samuel Purchas' Hakluytus Posthumus*, or *Purchas his Pilgrimes,* Contayning a History of the World, in Sea Voyages, & Lande Travels, by Englishmen and others is a huge compendium—published in 1625 and relating the travel accounts of many people, mainly English subjects.

happening that happen in brains that are less or not at all creative? Or does something qualitatively different happen? This is a very important distinction, which concerns many of the aspects of our lives; some differences are continuous—body weight, for example, varies from person to person in a gradual way, as does height, visual acuity—while others are discrete—such as biological sex, which, apart from the borderline case of intersex people, in the case of human beings, generally involves two mutually exclusive sexes. If creativity is continuous, then it can be studied just as we study intelligence, whereas if it is discontinuous, the approach used will have to change. Andreasen decides not to decide, or rather, to accept both visions: according to her, there are two types of creativity, the ordinary type, which can be studied with the tests we have already mentioned, and the extraordinary one, reserved for a few gifted individuals, violent and energetic like the creation of Kubla Kahn's palace, and which can be studied by means of neuroimaging, among other tools. And also through psychiatry, which is inextricably linked with neuroscience. Let us now begin a journey into the world of mental illness and the psychic structure of geniuses and creatives.

4.3   The 'Other Place'

We have seen how the personality of great creatives is made up of traits such as openness to new experiences, individualism, playfulness, a penchant for adventure, persistence, and so on. In particular, we have emphasised tolerance for ambiguity, that is, the ability to accept a world without certain answers and with blurred boundaries. The theme of nuance is also dear to the psychoanalytic approach, particularly in relation to the structure of the self: normally, in fact—according to this approach—the infant develops and acquires precisely that awareness of the existence of a distinction between their ego and the external world. Such 'ego boundaries' are, however, called into question in the case of certain mental pathologies, such as schizophrenia. It is as if—Andreasen tells us—the creative person lives 'on the edge of chaos,' and this fact entails the danger of losing the natural distinction between ego and external world. And even persistence has something obsessive, almost pathological about it: Michelangelo, for example, was known for his ability to work for very prolonged periods of time, without rest.

Interesting from this point of view is the case of Marvin Neil Simon, an American writer and playwright interviewed by Andreasen. According to the author, during his creative phase, he tends to slip into a state or condition of separation from reality, into what he calls "the other place," from the description of which it can be deduced that it is a dissociative state characterised by intense concentration, by detachment from everyday reality, in short, an altered state of consciousness that Andreasen compares to the states reached by the great mystics. And the American scholar goes so far as to suggest that highly creative people have brains that function differently, characterised by imaginative and perhaps fragmentary thoughts, but which flow quickly and, above all, without any censorial filter.

This dissociative state allows them to observe the world in a detached manner and without emotional involvement: "To others they may seem aloof, detached, or even coldhearted at times. To themselves, they often feel as if they are watching the rest of the world without others even knowing about it. This trait may seem to run counter to the flamboyance displayed by some creative individuals, who may appear to be seeking attention rather than invisibility. Nevertheless, many creative people, even when flamboyant, express a subjective sense that they can see into other people without being noticed. In a sense, they have an ability to spy on the universe."[6]As an example, the great Russian composer Pyotr Ilyich Tchaikovsky emphasised the purely unconscious and almost dreamlike nature of his creative process, pointing out its often violent and poignant characteristics. He also lamented how difficult it was to pick up the thread of one's creative thoughts once interrupted, how difficult it was then to regain contact with the source of origin, located just below the thread of consciousness. Of course, this statement should not lead us to think that the artist or the creative genius works only through inspiration; it is often the case that he or she works only because he or she has imposed it upon himself or herself, through discipline.

We have already seen the case of Poincaré, with his account of semi-conscious work. Rather, let us not forget the German organic chemist Friedrich Kekulé. who, according to his account, dreamt up the formula for benzene—although his account is full of obscure points that raise some doubts about its accuracy.

4.4   Inside the Enchanted Loom

And we finally come to the structure of the brain. This veritable scientific puzzle consists of two main parts, the telencephalon—itself subdivided into right and left hemispheres—and the cerebellum—located below the former.

---

[6] Andreasen, Nancy, Ibid., p. 39.

As for the two hemispheres, each is divided into four lobes, namely, the frontal lobe, temporal lobe, parietal lobe and occipital lobe. Now, we realise we are simplifying, but we can say that each lobe more or less performs some specialised function.[7] The frontal lobes serve for decision-making, the occipital lobes for vision, the temporal lobes for language and hearing, and the parietal lobes for language and spatial perception. The cerebral cortex contains between fourteen and sixteen billion neurons, while the cerebellum contains between fifty-five and seventy billion. And these neurons certainly do not sit idly by, but communicate with each other via synapses, forming what are called neural circuits. If this were not a metaphor, if neurons actually talked to each other, what we would hear would be a messy, annoying racket.

And what, then, is the brain? There are—is it any wonder?—many definitions and many theories, but we will focus on one that is in vogue, and that is also used by Andreasen, namely, the brain as SOS, Self-Organising System. We therefore return to Systems Theory, only this time, we restrict its application to the nervous sphere. Self-organisation is a key concept in

---

[7] You have probably heard over and over again about the specialisation of the two hemispheres—the right and the left—that make up our brain. Less well-known is the commissurotomy, a surgical procedure that consists of separating two parts of the body joined by a commissure—in practice, a joint. In the case of a cardiac valve commissurotomy, we are in the presence of a valvulotomy, while in neurosurgery, we have corpus callosotomy, a palliative procedure that consists of the complete bisection of the corpus callosum, the area that joins the two cerebral hemispheres. It is necessary in the case of very severe forms of epilepsy; by interrupting electrical communication, the intensity of epileptic attacks can be reduced, if not eliminated completely. It is not without consequences, this procedure: in fact, it leads to a high degree of mutual independence of the left hemisphere—the 'speaking hemisphere,' as the language functions are located there—and the right hemisphere—renamed the 'silent hemisphere.' It is as if the patients in question found two separate persons inside of themselves, with different functions. Thanks to this procedure, we have learned a great deal about the so-called lateralisation of brain functions. We know, for example, that the left hemisphere is in charge of logic, analysis, language, while the right hemisphere deals with visual and spatial reasoning, creation, imagination, intuition. To this, we must add that all of these findings remain controversial, that not all neuroscientists agree on the real degree of lateralisation of brain functions. The separation of left and right hemispheres has, however, served as an inspiration for several rather fanciful—and often mystical—speculations on the role that such lateralisation would play in human culture, with scholars ready to swear that the West would be characterised by the dominance of the left, logical and rational hemisphere, the East by the right, more 'holistic' one. Such is the case with Robert Ornstein, Stanford psychologist and author of a number of highly successful books on this subject, such as *The Psychology of Consciousness* and *The Evolution of Consciousness*. For our theme—genius—the work of Jan Ehrenwald, an American psychiatrist who has worked extensively on parapsychology and has attempted to interpret the great figures of Western culture in terms of hemispheric dominance, is certainly worth mentioning. Thus, Aristotle, Marx, and Freud would have lived an intellectual life in left-hemisphere dominance, while figures such as Plato, Nietzsche, Leonardo da Vinci, and Jung would have been characterised by right-hemispheric dominance. This line of interpretation is taken to the extreme by Ehrenwald; he goes so far as to claim, for example, that, in Beethoven's case, the odd-numbered symphonies—the Third, Fifth and Ninth—would have been written under the influence of the left hemisphere, while the even-numbered symphonies—such as the Sixth, i.e., the Pastoral—would have been written under the influence of the right hemisphere. See Ehrenwald, Jan, *Anatomy of Genius. Split Brains and Global Minds*, Human Sciences Press, New York 1984.

General Systems Theory and all related currents, such as Chaos Theory and Complexity Theory. It consists of the process whereby systems—especially living systems—composed of several parts spontaneously acquire their own structure and functions from relatively simple rules. Put simply, complex phenomena can arise spontaneously from simple elements, following a few principles; this is the case, using an example from zoology, when birds form spontaneously ordered flocks, or—from biology—when proteins fold to form articulated structures. And much more besides. In modern times, Kant—in relation to the formation of the solar system—and the idealist thinker Schelling both spoke of it. Then, we come to the twentieth century, and a field of research known as cybernetics, developed by Norbert Wiener and focused on the so-called feedback mechanisms, which govern all self-regulating phenomena—e.g., body temperature, metabolism, in short, all natural processes structured in such a way as to stimulate or inhibit themselves. In particular, the concept of self-regulation was first developed by W. Ross Ashby, in a book, *Design for a Brain*, published in 1952. In regard to more recent times, we must mention Heinz von Foerster, who speaks of 'second-order Cybernetics,' and Hermann Haken, who studies the phenomenon of self-organisation in a systematic manner and has created a special discipline, synergetics, relating to the mathematical basis of this concept.

As we have said, self-organising phenomena are found everywhere, including the brain. Consisting of a hundred billion strongly interconnected cells, the brain is able to produce very complex phenomena, such as thought, movement, decisions, and so on. What coordinates all of these cells so that these effects are achieved? According to systemicists, such as Haken, it would, in fact, be self-organising processes. A bit like walking: a gait consists of the momentary loss of balance and its immediate regaining. Similarly, the brain—or the other phenomena we have mentioned—would thrive in a kind of dynamic equilibrium in which chaos is what promotes order, through a constant process of reorganisation.

A typical example of self-organisation is found in perception, and in particular, in the phenomena of bistability of vision, in which an ambiguous image is perceived now in one way and now in another, precisely through a continuous oscillation between two different stable images—as in the famous case of the image of the young woman/old woman.[8] All of these phenomena can be interpreted as phenomena of nervous/cerebral/perceptual/cognitive self-organisation, according to Haken's model of synergetics.[9] And these

---

[8] https://www.illusionsindex.org/i/young-woman-or-old-woman.

[9] Haken, Hermann, *Synergetic Computers and Cognition* (2nd ed.), Springer, Berlin 2004.

phenomena would not be isolated cases: the entire brain would, according to this model, be a mass of feedback loops, of self-regulating processes whose properties would not be reducible to the sum of their parts. On the contrary, of all self-regulating systems, it would be the most complex, always there, generating thoughts, decisions and whatnot, often without even the need for external stimuli, but sheerly by its own power. And not just one type of thought, but two: conscious and unconscious. Now, the subject of the unconscious is rather complex and delicate, not least because it has been monopolised by psychoanalysis for so long; the idea, however, that there are thoughts that lie below the level of ordinary awareness seems to be beyond doubt.

So: the first thing at our disposal is a conscious, ordered thought, which comes into action when there is some specific physical or mental task to be performed, when we speak, when we tell a story, and so on. It is sequential thinking, with a beginning, a continuation and an end, in short, with a direction—it is the famous 'directionality' of conscious thought.

And then there is the unconscious thought, which 'flows' in the background of awareness, manifesting itself especially when the latter's influence is reduced, during sleep, or daydreaming. Non-linear thinking, apparently disordered, but actually with its own logic, at the root of normal and exceptional creativity. Apart from its therapeutic use in psychoanalysis, free association has had an important impact on artistic creativity—just think of William Faulkner—*The Sound and the Fury*—and James Joyce—*Ulysses, A Portrait of the Artist as a Young Man*. Interested in the neural basis of free association, Andreasen carried out a study of the phenomenon in question back in the 1990s using neuroimaging technologies in order to measure the blood flow in the brain to see which areas of that organ were activated during the activation of unconscious thought.[10] "The thinking in question," stresses Andreasen, "is based on a specific type of memory, the episodic memory. Originally studied by the well-known Estonian-Canadian cognitive neuroscientist Endel Tulving, episodic memory is our autobiographical memory, which concerns the stories we have experienced. It is connected to the passage of time, and concerns both the past—what we have experienced—and the future—what we fear or desire. Probably a uniquely human prerogative, episodic memory is contrasted with semantic memory, which concerns general notions—such as those we learn at school. Very important, episodic memory is at the root of our self-awareness and personal identity.

[10] Andreasen, N. C., D. S. O'Leary, T. Cizadlo, S. Arndt, K. Rezai, G. L. Watkins, L. L. Ponto, and R. D. Hichwa. *Remembering the Past: Two Facets of Episodic Memory Explored with Positron Emission Tomography*, American Journal of Psychiatry, 152, no. 11 (1995): 1576–85.

There would be no introspection without it. Nor would there be free association, although the type of episodic memory associated with it refers to 'deeper' events."

Andreasen's research has been based, in particular, on positron emission tomography (PET), in order to probe brain functioning from the blood flow in this or that area. In this field, studies are normally carried out by comparing the 'experimental condition'—i.e., the function being studied, e.g., language—with the 'control condition'—i.e., the brain of the experimental subject while he or she is at rest, with eyes closed. A problematic protocol, according to the researcher, since, in reality, when at rest, our brain is far from inactive; on the contrary, it often jumps very actively and, hopefully, pleasantly from one thought to another. Instead, Andreasen's study consisted in asking subjects to describe their day, starting from the morning wake-up, in what the researcher calls "focused episodic memory," comparing the PET results with those from another state, which she christened "random episodic silent thought" (REST)—a state induced by engaging the mind in a tedious semantic memory task, useful for seeing what happens to the free flow of semi-conscious thoughts when we do something without emotional involvement. Result: during the REST state, it was mainly the associative cortex that was activated, an area of the brain—distributed between the frontal, parietal and temporal lobes—responsible for collecting and weaving together information from the senses. Among the last parts of the brain to mature, the associative cortex is also typically human, being much more developed in us than in other mammals, which is why we can venture the hypothesis that so-called 'daydreaming' is the exclusive prerogative of our species. Of course, this is not the only finding of Andreasen's study; according to her research, both states studied were, in fact, connected to the activation of the associative area of the inferior frontal lobes and the precuneus, a parietal associative region. A sign that both states—'focused episodic memory' and 'random episodic silent thought'—were related to something very personal. In particular, previous studies have already shown that this inferior frontal region would be involved in the formation of individual value systems, guilt and awareness of one's social relationships. Damaging this area can produce antisocial behaviour, as in the famous case of Phineas Gage.[11] The moral of the story: free association has specific neural bases that are present in everyone, so creativity, which depends on it, is certainly to be

---

[11] Phineas P. Gage was an American labourer in the nineteenth century who was involved in a famous accident in which a metal pole pierced his skull, destroying his left frontal lobe. The man survived, but underwent a radical personality change, which accompanied him for the rest of his life.

considered an everyday phenomenon, rooted in the functioning of our own nervous system.

Does this also apply to exceptional creativity? That is, does it also have specific neural bases, perhaps—and this is a big question—different from those of ordinary creativity? Or, to put it another way, between ordinary creativity and exceptional creativity, is there—from a neurological point of view—a difference in degree or a qualitative difference? Does the highly creative person think like us—only better—or does he or she do it in a radically different way? We are not given to know at the moment, only to examine a few hypotheses. "It is certainly possible," says Andreasen, "that super-creatives function by means of neural processes that represent an augmented or more complex version of those that fuel ordinary creativity." Think of writers, for example. Great writers are, contrary to popular belief, on average very disciplined; they sit down and write, whether they feel like it or not. It is a state of intense concentration, which cannot last very long. And, from time to time, this state of concentration leads to insight, also described above through personal accounts, such as Coleridge's. Such descriptions represent introspective accounts of what we have called 'extraordinary creativity.' Often, the composition of a text or the making of a discovery occurs precisely in a flash, obtained while the subject is immersed in a semi-honest state. The neurological hypothesis put forward by Andreasen is more or less the following: in great creatives—but don't feel down, it could happen to you, too—the REST state would allow for the free flow of ideas, images, and so on, in a process that starts with *disorganisation,* and then arrives at cerebral *self-organisation,* and ends with an original product, be it a song, a poem or a theory. In essence, the different parts of the associative cortex would send messages back and forth to each other, not for the purpose of integrating them with motor or sensory input, but as a simple reciprocal response. Such associations would occur freely, without censorship, without the intervention of the Freudian 'reality principle.' The state of disorganisation—which can persist for several hours—allows for the collision of ideas, images and concepts, and the emergence of the new. It would therefore be a rare type of creativity and, above all, qualitatively different from everyday creativity—i.e., the creativity we employ, for example, to compose an email, prepare a shopping list or solve a problem at home or at work. Disorganisation, the free flow of ideas, 'cognitive precipice' leading to sudden reorganisation.

Those gifted with extraordinary creativity would therefore have brains capable of creating free associations of this kind more easily. Such brains could be equipped with an associative cortex richer in connections among its different parts, or could even have a different kind of connection. The latter would be what really distinguishes geniuses from us mere mortals, but there would be a price to pay, because a certain degree of psychological vulnerability would also be associated with this characteristic. In some cases, extreme vulnerability: think of the schizophrenia of the mathematical genius John Nash, who believed he had been recruited by aliens in order to save the world, or Vincent van Gogh, who died by suicide at the age of 37 after suffering all his life from manic-depressive episodes.

4.5   The Dilemma of a Cure

Far from limiting herself to collecting anecdotal accounts, Andreasen has also been conducting experimental research into creativity and psychic pathologies since the 1980s. In particular, one of her studies examined participants in a well-known creative writing programme, the University of Iowa Writers' Workshop—founded in the 1940s by the poet and novelist Paul Engle. Highly prestigious, the programme in question was the first in the US to award a doctorate in creative writing; among its students—and lecturers—we find the cream of American literature, including—just to name a few—Kurt Vonnegut, John Irving, Robert Lowell, Anthony Burgess, Mark Halperin, Mark Strand, Robert Coover, Philip Roth, John Cheever, Jori Graham and Gail Godwin. The study, published in 1987, examined thirty writers, thirty other outsiders—as a control group—and first-degree relatives of both groups.[12] Most of the writers were found to suffer from mood disorders that could be classified under the categories of bipolar disorder or unipolar depression—most had received psychiatric treatment for these problems. An interesting aspect emerges from this research. As is well known, mood disorders tend to be characterised by more or less short periods of high or very low mood, interspersed with longer periods of normal mood. The writers were interviewed during a period of normal mood, so as to allow them to look at their condition with a certain detachment and, above all, to describe the influence of their disorder on creativity as objectively as possible. And the answers were almost unanimous: manic or depressive states are an impediment to creative work. From the point of view of mood, such negative states do not seem to be permanent, and in some cases, they are a source of inspiration for subsequent work. In addition, Andreasen's research also revealed the presence of such disorders in the relatives of the writers examined, thus highlighting their probable genetic origin. And what about schizophrenia? According to

---

[12] Andreasen, Nancy C., *Creativity and mental illness: Prevalence rates in writers and their first-degree relatives,* The American Journal of Psychiatry, 144(10), 1288–1292, 1987.

the researcher, the association between artistic creativity and mood disorders is quite strong—as has also been shown by several studies since hers—as is the *absence of* schizophrenia. From this, it can be deduced that artistic— or poetic, or literary—creation would be incompatible with such a serious mental pathology, which drives those who have it to isolation and, above all, to cognitive disorganisation. This does not detract from the fact that, in other areas of life, schizophrenia and creation can coexist—especially where creation occurs through 'flashes of genius' or through the exploration of 'sensations' and 'intuitions.' In this regard, there is no shortage of anecdotes. In addition to the aforementioned Nash—and his son, who also suffered from schizophrenia—we recall Bertrand Russell; in his case, we have no evidence that he suffered from it, but we do know for certain that his family tree had several cases of diagnoses of schizophrenia. And other prominent scientists— mainly mathematicians and physicists—have had some relationship with this condition. Think, for example, of Isaac Newton—about whom, mind you, we possess only anecdotes, not least because the scientist lived before psychiatry was established as a science. Suspicious to a degree that we would today describe as pathological, Newton cultivated interests—such as astrology and alchemy—that were considered somewhat bizarre even at the time; he lived much of his life in solitude, never marrying. In addition, at the age of 51, in 1693, the scientist suffered a nervous breakdown, peppered with persecution mania. According to Andreasen, Albert Einstein may also have manifested at least some schizotypal traits—a mental disorder characterised by social and interpersonal deficits and bizarre behaviour. His son with Mileva Maric also suffered from schizophrenia.

And it is here that we find—according to Andreasen—the link between creativity and mental illness. In particular, it has been shown that the extraordinary creative process would be potentially dangerous, constituted as it is by free associations that haunt the brain within unconscious or semi-conscious mental states, by thoughts that must momentarily disorganise themselves and then reorganise. And this process would be precisely similar to what happens in the nervous system of people suffering from schizophrenia, depression and manic states. The famous Swiss psychiatrist Eugen Bleuler, who coined the term 'schizophrenia,' believes that the most typical feature of this pathology is precisely the loosening of mental associations. "Therefore, when such associations recompose themselves, reorganising themselves," reasons Andreasen, "we have creativity, when they fail—or something goes wrong, for example the result obtained is completely detached from reality—we have schizophrenia."

Andreasen's studies then consider another very interesting aspect of the relationship between creativity and mental illness, namely, the effects that treating the latter might have on the former. Does treating one's psychological discomfort and psychiatric disorders—with psychotherapy or psychotropic drugs—eliminate creativity? Does it leave it intact? Or does it increase it? There is much confusion on the subject, even among those directly involved. There are those who fear that, by being cured, they will lose the supposed advantages that a manic condition can bring—sky-high creativity, energy, and so on. And it must also be said that some creatives have literally romanticised their sufferings, ending up seeing in them a source of inspiration. Nevertheless, treatment is often necessary, and many creatives—once cured and stabilised—feel better and would certainly not go back. Let us also not forget the risk of suicide that an untreated mental illness entails: van Gogh, Virginia Woolf, Hemingway, and many others are there to testify to this.

What does medical science say about the relationship between creativity and psychotropic drugs? In this regard, we have a famous study, published in 1979 by the Danish psychiatrist Mogens Schou, who was one of the first to study the effects of lithium carbonate in the treatment of bipolar disorder.[13] Schou examined a group of artists suffering from this disorder and measured their productivity, as well as the artistic quality of their production. Schou then divided the subjects into three groups according to the type of response they had to lithium treatment. The first group consisted of people suffering from a severe form of bipolar disorder and who, thanks to lithium, saw an improvement in their condition and an increase in their artistic productivity. The second group consisted of people who experienced no alteration in their creativity, either positive or negative, to the extent that Schou suspected that these subjects were not really taking the therapy. The third group saw a decrease in their creativity; these were mainly subjects who based their artistic productivity on manic peaks.

And the usefulness of the treatments is also confirmed by Andreasen's study in the Writers' Workshop: the majority of those interviewed confirmed their inability to write during periods of depression. In spite of this, research on the relationship between creativity and the effects of psychotropic drugs is rather scarce, precisely because it requires the availability—not always easy to obtain—of scientists and artists suffering from such mental illnesses. So, we have to make do with anecdotal reports.

Having said that, let us try to answer—from a neuroscientific perspective—the following question: is it possible, if not to become geniuses, at

---

[13] Schou, Mogens, *Artistic Productivity and Lithium Prophylaxis in Manic-Depressive Illness*, British Journal of Psychiatry, 135 (1979): 97–103.

least to improve one's mental abilities, just as we improve our physical ones? Without a doubt, Andreasen tells us, thanks to the notion—which we have already encountered—of neural plasticity, a notion stemming mainly from the work of psychiatrist and Nobel Prize winner Eric Kandel. Having said that, to get down to practicalities, what does the American neuroscientist and psychiatrist advise us to do? She suggests some simple exercises that could help us. In particular, Andreasen suggests choosing a new and unfamiliar area of knowledge and studying it in depth; practising meditation, so as to open the mind to thoughts of all kinds, even unusual ones, learning self-discipline and perhaps reaching Simon's 'other place'; practising observation and description, for example, by choosing a specific theme, such as birds, works of art, landscapes, or something else; finally, practising visualisation and imagination, literally trying to 'see' with the mind's eye what we are reading about. We may not become geniuses by doing this, but we will certainly improve our creativity.

4.6   If-And-Then

You may have looked up at the sky and glimpsed particular shapes in the clouds, or seen 'faces' in the patches on the walls. This is a perceptual phenomenon known as 'pareidolia,' and testifies to the existence in us of an innate tendency to attribute—perhaps as a precautionary measure, i.e., to prevent predators from attacking—intentions to things that lack them. And pareidolia represents just one case of a more general human ability, namely, the ability to recognise patterns and regularities in the natural world. Not only that: the whole of human scientific production would be the fruit of such an ability. Simon Baron-Cohen, a well-known British clinical psychologist and director of the Autism Research Centre at the University of Cambridge, put forward this hypothesis—which is also useful for us to better understand what distinguishes geniuses from us mere mortals. Not only that, but, according to the scholar, this tendency is also connected to a specific human condition that is still poorly understood, autism.[14] The search for consistent and repeated patterns is typical, however, to varying degrees, of all human beings, and this is especially noticeable in children, true little natural scientists. And, according to Baron-Cohen, some minds function in an accelerated mode, and are thus able to produce revolutionary inventions and great discoveries.

---

[14] Baron-Cohen, Simon, *The Pattern Seekers—How Autism Drives Human Invention*, Basic Books, New York 2020.

Like Simonton, Baron-Cohen believes that our mind is the fruit of evolution, which has endowed it with an internal mechanism aimed at the discovery/construction of hypothetical patterns based precisely on the if-and-then model, allowing us to identify repeated patterns. Particularly, in some minds, this mechanism would be 'tuned' to the maximum, allowing them to focus on the most minute details, even going so far as to take pleasure in discovering how a certain mechanism, of whatever kind, works. Autistics, natural scientists and, in general, all 'hyper-systemizers'[15] would have in common such a mechanism in its accelerated version.[16] Baron-Choen's 'Systemizing Mechanism' has four steps; to be precise, the subject/mind/brain begins by asking itself a question—of the 'why', 'how', 'what' or 'when' type—concerning the external or internal world. The second step consists in answering the proposed question by means of a hypothesis concerning a causal relationship of the 'if-and-then' type. The third step consists in repeatedly testing this hypothetical scheme, dozens or, if need be, hundreds of times. Finally, the fourth step consists in modifying, adapting the hypothetical scheme, testing it again as many times as needed. "Depending on the field," the scholar points out, "we are in the presence of a discovery or an invention."

These hyper-systemizers might choose to go into science, engineering, or sports, or perhaps playing a musical instrument or any discipline that requires a high degree of precision. In terms of timing, according to Baron-Cohen, this mechanism would work in our brains for a minimum of seventy to a maximum of one hundred thousand years. It was the nineteenth century English mathematician George Boole, to whom Baron-Cohen gives credit for this, who first described the Systemizing Mechanism from a logical point of view; for this reason, the psychologist has renamed this uniquely human characteristic the 'Boolean Mind'. In short, our species is said to have an at least partly innate mechanism for discovering patterns in its environment.

In order to test their if-and-then hypotheses, humans would use three distinct procedures, namely, observation, experimentation and model development. The Systemizing Mechanism can be tuned to different levels of intensity; in the average individual, it is tuned to a medium level, while in a limited number of people, the geniuses, it is tuned to very high levels of activity. This would be a largely genetic-based mechanism, so it would tend

---

[15] When we identify an if-and-then pattern, we are discovering a system, hence the term "hyper-systemising".

[16] The Swedish botanist and zoologist Carl Linnaeus must have been a hyper-systematiser, in the sense that he dedicated his life to systematising precisely all of the living forms that came his way.

to be transmitted through generations and, in the eternal debate between nature and nurture, would tip the balance towards the former.

And those with a high level of activation of this mechanism would be able to absorb a huge amount of information and organise it into if-and-then-type patterns, thus tracing it in their environment and the world at large. The evidence that hominids—including our closest cousins, the Neanderthalians -, primates and other animals can effectively trace if-and-then patterns is rather scant, according to Baron-Cohen; the Systemizing Mechanism might, in fact, be a trait exclusive to modern humans.

Alongside this mechanism, we find what the English scholar calls the Empathy Circuit, which includes a 'theory of other people's minds', and which is at the root of our ability to intuit other people's intentions. In principle, if the Systemizing Mechanism is tuned to the maximum, the Circuit of Empathy is tuned to the minimum, and vice versa.

According to the work of Simon Baron-Cohen, it is possible—by looking at the level of activation of the Systemizing Mechanism and the Empathy Circuit—to classify the brains of human beings into five types, distributed differently in the general population. The reflections in question are the result of an extensive study carried out by Baron-Cohen himself and colleagues, dubbed the UK Brain Types Study and involving no less than six hundred thousand volunteers, with the aim of measuring their empathy and systemising traits.[17] The volunteers were given questionnaires concerning their cognitive functioning.[18] According to the British researcher and his colleagues, the afore-mentioned five human brain types, from the point of view of the mechanisms in question, could be classified as such: we have Type B (balanced), in which the two mechanisms are precisely balanced; Type E, in which empathy is tuned at high levels, and systemization at low levels; Type S, which is the opposite; and finally we have Type E Extreme and Type S Extreme.

Only a small fraction of the general population, about three per cent, belongs to Type S Extreme, a human type that we could label 'hyper-systemizers'. And, according to Baron-Cohen, this characteristic is a prerequisite for the human capacity to invent. It would certainly be more widespread, were it not for its connection with little or no skill in social relations. In short, the relationship between the Empathy Circuit and the Systemizing Mechanism is, to all intents and purposes, a zero-sum game. And, if we were to take

---

[17] https://yourbraintype.com/.

[18] Greenberg, David, et al., *Testing the Empathizing Systemizing (E-S) theory of sex differences and the Extreme Male Brain (EMB) theory of autism in more than half a million people*, Proceedings of the National Academy of Sciences 115(48), 12,152–12,157, 2018.

a look at modern great inventors, we would find that, in them, the System-izing Mechanism is tuned to the maximum, that their brains are Type S Extreme. There is no doubt, then, according to Simon Baron-Cohen: creative abilities are largely genetic in origin.

Most of us look for solutions to problems that are good enough. In this regard, Nobel Prize winner and psychologist Herbert Simon calls this trait 'satisficing'; in essence, we ordinary people would be 'satisficers'[19]; extreme systemisers on the other hand—whom we could also call 'maximisers'—seek the optimal solution. The Extreme S-type of brain basically functions differ-ently from the norm, and this is because its operating system is different. For example, Thomas Alva Edison could not help but "systemize." Appar-ently, the inventor once told a journalist: "I have not failed; I have simply found ten thousand ways that won't work."[20] And after all, Edison, as a good hyper-systemizer, lacked social skills, neglected family obligations and devoted himself obsessively to work, reaching or even exceeding eighteen hours a day for weeks on end, even neglecting personal hygiene and meals. In essence, the if-and-then algorithm, gifted to humanity by evolution, was tuned to the highest level in Edison's brain. And, while we're on the subject of great inventors, Bill Gates, talking about himself in his twenties says: "I was a zealot. I didn't believe in weekends. I didn't believe in vacations."[21] Another example of a hyper-systemizer would be, for Baron-Cohen, the famous clas-sical pianist Glenn Gould, who possessed an excellent musical memory and practised his art according to an obsessive routine, which included control-ling all aspects of his life, e.g., the exact temperature of the room in which he practised, the use of his special chair that allowed him to sit lower than the keyboard. A repetitive behaviour that characterised him—perhaps a physical manifestation of the if-and-then mechanism—was his rocking back and forth while playing the piano. Some have speculated that Gould was autistic, although a formal diagnosis has never been made.[22] The identifica-tion of hyper-systemizing traits superimposed on autistic traits could concern a large number of personalities belonging to the most diverse fields: e.g., Henry Cavendish and Albert Einstein for physics, Hans Christian Andersen for literature, Ludwig Wittgenstein for philosophy, Andy Warhol for art, although—Baron-Cohen points out—making such diagnoses on the basis

---

[19] Simon Herbert, *Rational choice and the structure of the environment*, Psychological Review 63(2), 129–138, 1956.

[20] Quoted in Elkhorne, J. L., *Edison—The fabulous drone*, in 73 Vol. XLVI, No. 3 (March 1967), p. 52, 1967.

[21] Galanes, Philip, *The mind meld of Bill Gates and Steven Pinker*, New York Times, January 27, 2018.

[22] McLaren, Leah, *Was Glenn Gould autistic?*, Globe and Mail, February 1, 2000.

of more or less incomplete biographical accounts is always a problematic methodology. And, from a point of view of the British scholar's approach, being a hyper-systemizer does not automatically mean being autistic, so the link between the two traits is not absolute—although being a hyper-systemizer does increase the possibility of possessing autistic traits. In any case, here it is, for Baron-Cohen, the secret ingredient of genius, an ingredient, however, not available to everyone, but only to those who possess the right genetic make-up: the systemizing mechanism, a gift of evolution.

4.7 An Expanding Sector

In recent years, research on the neurological basis of creativity has multiplied and flourished, introducing many innovative ideas.[23] Thus, according to Alice W. Flaherty, an American neurologist and psychiatrist, human creativity would depend on a high level of motivation, but not too high: the human brain would, in fact, exert a rather tight control over drives, and the balancing of opposing motivations. The brain is endowed with two central motivational systems, one for approach and the other for avoidance, and both would exert a noticeable effect on creativity. In particular, the former, the dopaminergic system—based on the neurotransmitter dopamine—stimulates goal-driven motivation. Serotonin, on the other hand, increases fear-driven avoidance motivation. The dominant left hemisphere helps in the tracing of familiar patterns, while the right stimulates avoidance motivation, but also the identification of novelty. In the creative sphere, this dual system coordinates the incorporation of new perceptions into new patterns.[24] According to the Canadian psychologist Michael C. Corballis, contrary to a more traditional approach, creativity would depend on brain circuits that extend throughout the entire brain, including the default mode network,[25] and there would therefore be no polarity between the right and left hemispheres, an idea

---

[23] For those who would like to keep abreast of the latest developments in neuroscience in this specific field, we recommend: Jung, Rex E. & Vartanian, Oshin (Eds.), *The Cambridge handbook of the neuroscience of creativity*, Cambridge University Press, Cambridge 2018.

[24] Flaherty, Alice W. (2018). Homeostasis and the control of creative drive. In R. E. Jung & O. Vartanian (Eds.), *The Cambridge handbook of the neuroscience of creativity*, Cambridge University Press, Cambridge 2018, pp. 19–49.

[25] The default mode network (DMN) is a large brain network composed mainly of the medial prefrontal cortex, the posterior cingulate cortex and the angular gyrus; it is usually activated when the mind is not paying attention to the outside world and is daydreaming.

that had seemed a given.[26] Finally, we would like to mention the interesting work of Claudia Garcia-Vega and Vincent Walsh, who have dealt with a phenomenon only partly related to that of genius, namely, polymathy. As is well known, it consists of the ability of some people to deal in depth with the most disparate topics, even managing to produce original or highly original contributions. A classic example is that of Leonardo da Vinci, artist, scientist and inventor. In their search for a neural correlate of polymathy, Vega and Walsh believe they have found it in the aforementioned default mode network. Indeed, it, together with related psychological phenomena—such as daydreaming—is not specific to a precise cultural domain, but is indeed 'interdisciplinary.'[27]

---

[26] Corballis, Michael C., *Laterality and Creativity: A False Trail?* In: R. E. Jung & O. Vartanian (Eds.), Op. Cit. pp. 50—57.
[27] Garcia-Vega, Claudia & Walsh, Vincent, *Polymathy: The Resurrection of Renaissance Man and the Renaissance Brain*, in: Jung, Rex E. & Vartanian, Oshin (Eds.), Op. Cit., pp. 528–539.

5

The Contradictory Heart of Genius

5.1 Nature or Nurture?

Let us now enter into the heart of our work, and try to summarise the essence of the research—a research that has been going on for about a century and a half—on the subject of genius. In the course of this journey, we have indeed come to the conclusion that there is no single answer to the question of what genius is, that many contradictory statements are simultaneously true. Let us begin with the first question: does genius depend on nature or nurture? It is a question that goes back at least as far as Galton, who seeks precisely to prove that the needle of the scales leans towards nature. From a historical point of view, the opposing concepts of 'nature' and 'nurture' date back to William Shakespeare, and, in particular, to *The Tempest*; it is Prospero, the sorcerer protagonist of the play, who speaks about it. Specifically, the character refers to Caliban—a creature only half human—as "A devil, a born devil, on whose nature/Nurture can never stick"; in the eternal debate, the magician thus seems to side with nature. As we have seen, this dichotomy was made famous, however, by Galton himself, a great advocate of the idea that one is born a genius, not made one—so much so that he entitled his 1869 book *Hereditary Genius*. And so much so that he devoted a fair amount of his time to compiling lists of eminent people associated with their respective family trees—scientists, men of letters, poets, but also leaders, jurists,

© The Author(s), under exclusive license to Springer Nature
Switzerland AG 2023
R. Manzocco, *Genius*, Springer Praxis Books,
https://doi.org/10.1007/978-3-031-27092-5_5

and so on. The Bach family, for instance, comprises at least twenty eminent musicians—including Johann Sebastian Bach's sons Carl Philipp Emanuel, Johann Christian, Wilhelm Friedemann and Johann Christoph Friedrich. Not that all the Bachs were musicians of genius level, but eminence in that family was quite widespread. Other eminent families included in Galton's lists include the Bernoullis—mostly mathematicians-, the Cassinis—several of them astronomers-, the Herschels—again, astronomers, except for John Frederick William, who was also a physicist and chemist. Galton then goes on to include in his genealogical researches his own family and, in particular, the Darwin family, starting with his grandfather Erasmus. The father of evolutionism appreciated the work, shared its genetic determinism and agreed with Galton that education was a negligible factor, thus giving priority to innate qualities. How would you like to be included in Galton's list? Apparently, Alphonse Pyramus de Candolle (1806–1893), a well-known Swiss botanist and son of the equally eminent botanist Augustin Pyramus de Candolle, did not like it at all, to the point that he set to work precisely to prove the opposite thesis: the scientist thus collected data—published in 1873 in the book *Histoire des sciences et des savants depuis deux siècles*—that would show that genius arises in specific countries and at specific times in history, a sign of its essentially cultural nature. The following year, Galton published another book, *English Men of Science: Their Nature and Nurture,* in which, although he does not renounce siding with nature, he also concedes some space to nurture.

Despite all of the critical aspects of Galton's work, he nevertheless contributed to the emergence of a very fruitful and, for us, very useful field of research: behavioural genetics. This field of research studies the way in which individual differences—especially character differences, but not only these—are transmitted from one generation to the next. And this discipline has given us a very useful analytical tool, namely, the study of twins. As you may know, there are two types of twins, identical twins and fraternal twins. The former—called monozygotic—come from the same fertilised ovum that accidentally split in two; physically identical or nearly so, they often display unusually similar intellectual capacities and psychological inclinations. Seconds—called heterozygotes—come from different fertilised eggs, do not—unlike monozygotes—share identical DNA and are as similar to each other as two brothers born at different times could be. The study of identical and fraternal twins is very useful precisely because it allows us to separate what depends on genes from what depends on the environment, especially if we consider identical twins adopted from different families: in this case, in fact, the genetics are the same, while the family environment is different. And analyses carried out by behavioural genetics on subjects of this type have revealed how, in fact, intellectual abilities, vocational inclinations and even specific tics and minor obsessions would be hereditary in nature. Let us now turn to the analysis of a concrete case, that of William James, and a specific characteristic commonly attributed to geniuses, openness. If subjected to today's psychological tests, James would be found to possess a very broad openness: an aspiring artist as a young man, he even began an apprenticeship at the studio of a famous American painter, William Morris Hunt. He quickly switched to the sciences, being admitted to Harvard Medical School, and later suspending his studies to participate in an expedition to Brazil. He then travelled to Germany, coming into contact with philosophy and psychology, to which he later devoted himself, writing classic works such as *Principles of Psychology* and *Varieties of Religious Experience*, a masterpiece of the psychology of religion. He also dealt with paranormal phenomena, joining the Theosophical Society and founding the American Society for Psychical Research. Not bad for an open mind. Now, undoubtedly, his family background—a wealthy and open-minded father—was certainly helpful to William James, and would seem to explain his personal events in terms of nurture, were it not for the fact that behavioural genetics has long since highlighted the essentially hereditary, genetic nature of openness. In short, it seems that modern science has done justice to Galton.

Not so fast, though. There are, in fact, some problems with this view of things. Noted behavioural genetics scholar David T. Lykken points out in his research that homozygotic twins are *too* similar. That is to say, they—even if bred separately—show a very strong similarity in traits that appear to be non-inherited. Or, at least, that do not appear to be hereditary based on studies of heterozygotic twins. Now, if homozygotic twins have a certain trait in common, we would expect that trait to appear in attenuated, halved heterozygotic twins, as happens between siblings. Lykken studies 'social potency,' i.e., the ability to command or influence others. According to our reasoning, if homozygotic twins fully share this trait, statistically heterozygotic twins should share it less often, but still share it. Well, Lykken points out instead that, in the case of social potency, while homozygotic twins share it fully, heterozygotic twins statistically do not share it as often as they should; from the point of view of numbers, heterozygotic twins are no different from two individuals taken at random in the general population. The answer that the

American scholar gives refers to a very interesting concept, that of 'emergenesis.'[1,2,3,4] It consists of the idea that, while certain genetic traits depend on additive factors, other traits depend on multiplicative factors. To put it more clearly: height is a trait based on additive factors, i.e., on the contribution of different genes, each of which contributes independently to the trait in question. Other traits, such as genius, on the other hand, depend on multiplicative factors; in essence, in order for a certain trait to 'stand out,' certain genes must be present collectively, i.e., all of them must be present, bar none. It is not enough to have almost all of them: it takes the presence of each and every one of them. Among other things, this would explain why homozygotic twins exhibit traits that heterozygotic twins do not: the former would, in fact, have the complete set of genes necessary for a certain trait, the latter would not. For Lykken, genius would be just like that, and this theory—which, at the moment, is only speculative, though certainly fascinating—would explain why, in certain cases, we find creative geniuses without ancestors and equally brilliant descendants. Think of the great mathematical genius Karl Friedrich Gauss: the son of a bricklayer and a peasant woman, he had only one son, who never managed even to approach his father's heights. Here, Gauss, for Lykken, would be a clear case of emergenesis.

But what about nurture at this point? Let us also try somewhat to defend this second point of view, which we have hitherto mistreated. And let us do so through the Darwin family. What do the genes possessed by this genius family matter? The Darwins were able to take advantage of economic and educational opportunities precluded to the majority of Her Majesty's citizens; Charles's father, Robert, was a well-known doctor and an able investor; his father Erasmus ensured that his son could study at prestigious universities in the UK and Europe. Robert later married the daughter of a wealthy industrialist friend of the family, had six children and smoothly sent Charles and his elder brother Erasmus to study at prestigious universities, where the two found teachers and mentors of the highest calibre, including, in Charles's case, the botanist John Stevens Henslow, who recommended his protégé as a naturalist for the crucial voyage on the Beagle. Had Darwin's father not allowed his son to participate in such an undertaking, it is likely that the history

---

[1] Lykken, David T., *Research with twins: the concept of emergenesis*, Psychophysiology 19, 361–73, 1982.

[2] Lykken, David T. et al., *Emergenesis: genetic traits that do not run in families*, American Psychologist, 47, 1565–77, 1992.

[3] Lykken, David T. et al., *Heritability of interests: a twin study*, Journal of Applied Psychology, 78, 649–61, 1993.

[4] Lykken, David T., *The genetics of genius*, in: Steptoe, Andrew (Ed.), *Genius and the Mind. Studies of Creativity and Temperament*, Oxford University Press, Oxford 1998, p. 15–37.

of evolutionism would have been different. On the basis of this portrait, it would seem that Darwin was a product of circumstance rather than lucky genes. And let it not be thought that things are very different for other genius scientists. According to Galton's research, for example, a very high percentage of the scientists he interviewed came from upper-class families—doctors, civil servants, lawyers, bankers. And Anne Roe's later studies showed that there had been a change over the intervening decades, in the sense that, at the time of her research, the bulk of eminent scientists came from professional families, engineers, and so on. This is as far as the economic starting point is concerned. As for the educational one, Galton already pointed out that most eminent scientists came from elite universities. And things have not changed much over time. In this regard, in the 1970s, the sociologist of science Harriet Zuckerman reiterated the elitist nature of contemporary science education, with the bulk of the successful scientists she surveyed coming from America's most exclusive universities.[5] Have we succeeded in establishing the primacy of nurture over nature? We believe not, although the latter certainly plays a very important role in all of this. But remember what we repeat at regular intervals in this book: there are no rigid rules, i.e., it is always possible to find exceptions. This is precisely why there is no shortage of cases of people who, despite genius ancestors or relatives, are themselves lacking in eminence, as well as geniuses who have no high-ranking relatives—a fact that we have tried to explain with emergenesis, but without arriving at anything certain. So, let us go back to the Darwins, and cite the case of Charles's elder brother, Erasmus, who, despite having grown up in the same family environment, despite having followed a similar schooling and even having worked with a well-known mentor at Cambridge—James Cummings, a chemist—and despite having exchanged opinions at length with his younger brother, produced nothing significant. So much so that his father literally granted him an annuity at the age of twenty-six. Things can also go the other way, however: this is the case with Newton, whom Galton tried hard to link to an eminent genealogy, but failed. The emergenesis hypothesis aside, the ancestors and descendants—collateral, since Isaac had no children—of the English physicist are, in fact, nothing exceptional. And Newton is certainly not the only case of genius emerging from a less than eminent family background: think of Michelangelo, or none other than Leonardo da Vinci.

At play in the field of nurture are so-called "diversifying experiences." To understand what this is all about, let us take a step back and brush up on some sociology. Far from being a neutral approach, sociology, in fact, starts

[5] Zuckerman, Harriet, *Scientific Elite*, Free Press, New York 1977.

from strong assumptions, namely, the idea that the individual is an epiphenomenon. In essence, we individuals would be the product of impersonal social forces, a process referred to in sociology as 'socialisation.' Nothing to do with making friends, socialisation consists precisely of the social forces that imprint values, rules, and identity itself on the individual's 'blank slate'. We are all therefore subjected to similar social pressures—relating to everyday behaviour, rules of good manners, what is desirable in life and what is not: if this were not the case, the social order could not exist. Which also explains why anti-conformists represent a minority: even in societies like ours, where anti-conformism is praised, one still ends up conforming to the general rules, otherwise society could not function. This is not the case for many creative geniuses, however, who, in the course of their lives—especially in the early stages—would have encountered precisely 'diversifying experiences,' in essence unusual, often traumatic situations, which would also have substantially altered their socialisation process. Think of Ludwig van Beethoven, whose father was an alcoholic and also a violent man, while his mother died when he was very young. Or Descartes: a rather sickly boy, he lost his mother at the age of one and was raised by relatives. Or Dostoyevsky, who suffered from childhood, and throughout his life, from epileptic attacks. Or Freud's destitute childhood, Byron's deformity, Edison's enormous difficulties at school. And here, for example, is a review—incomplete, but significant—of scientists who lost one or both parents in the first ten years of their lives: d'Alembert, Berzelius, Boyle, Buffon, Cavendish, Copernicus, Darwin, Eddington, Fleming, Fourier, Fulton, Humboldt, Huygens, Jenner, Kelvin, Lavoisier, Lobachevski, Maxwell, Newton, Paracelsus, Pascal, Priestley, Quetelet. And philosophers: Confucius, Descartes, Hobbes, Hume, Leibniz, Mencius, Montesquieu, Nietzsche, Rousseau, Bertrand Russell, Shankara, Sartre, Spinoza, Swedenborg, Voltaire. Writers: Baudelaire, the Bronte sisters, Byron, Camus, Coleridge, Conrad, A. Cowley, Dante, DeQuincey, Donne, Emerson, Gibbon, Gorky, the Brothers Grimm, Hawthorne, Holderlin, Keats, Lermontov, E. B. Lytton, Mallarme, Maugham, Moliere, Montaigne, Neruda, Poe, Racine, Solzhenitsyn, Stendhal, Swift, Thackeray, Tolstoy, Wordsworth, Zola. Artists: Canova, Delacroix, Masaccio, Michelangelo, Munch, Raphael, Rubens. Composers: J. S. Bach, Puccini, Sibelius, Wagner. And we could go on and on. Some hypotheses regarding diversifying experiences are possible. First, we have the bereavement syndrome hypothesis: the loss of one or both parents would drive the budding genius to seek a form of emotional compensation in the achievement of their creative goals. The second hypothesis concerns the fact that such a loss would strengthen the genius in question to such an extent that they would be able to overcome all of the adversities on their path to eminence. The third hypothesis—and this is the one favoured by sociology—posits that such trauma would push its victims towards unusual paths of psychological development.

What is the consequence of such childhoods? Probably the fact—and this is precisely the aforementioned sociological interpretation—that they weaken the rigidity of social rules on an individual level, allowing those who experience such things to think in unconventional ways. Loss of parents, poverty or economic instability, geographical displacement, physical or mental problems can shape lives in unusual ways, and can sometimes produce creative geniuses with entirely new ways of thinking, which follow an existential path divergent from the norm. From this point of view, Simonton's analysis of Nobel Prize winners is interesting, according to which scientists would tend to come from stable home backgrounds, while literati would come from unstable or at least less conventional situations. For example, physicists would have academic fathers in 28 per cent of cases, chemists in 11 per cent, and writers in 6 per cent. Only 2 per cent of physicists lost their fathers as children, while this happened to 11 per cent of chemists and 17 per cent of writers. If we put together being orphaned by at least one parent, financial problems or extreme poverty, the percentage of writers who have experienced at least one of these diversifying experiences rises to 30 per cent. On average, scientists also tend to have followed a more comprehensive academic path than writers—but it could not be otherwise, especially nowadays, given the level of formal training required in contemporary research. Now, it seems that these diversifying experiences speak in favour of the primacy of nurture, but is this really the case? Certainly, losing parents is not written in the genes. And negative proof of this comes to us from the Termites, the Terman gifted children, who, in addition to being physically healthy, all came from stable and happy family situations; in short, perhaps we have to conclude that having too happy a childhood harms creativity?

What is also certain is that the debate is still very much open, and that a single 'genius gene' probably does not exist: reason asks us to suspend judgement for now, and attribute genius to a complex interweaving of genetic and environmental factors. In short, if we are looking for clear-cut answers—nature or nurture—we are dealing with the wrong subject.

5.2   The Price to Pay

Here, we come to the second question: madness or sanity? There is, in fact, no one who is not familiar with the cliché of the mad genius, the person who is talented but tormented by one or more mental disorders. A cliché that dates back to Antiquity, since Aristotle is said to have stated that "Those who have become eminent in philosophy, politics, poetry, and the arts have all had tendencies towards melancholia." A similar opinion is traditionally attributed to Seneca, while, centuries later, Shakespeare wrote, in *A Midsummer Night's Dream*, that "The lunatic, the lover, and the poet/Are of imagination all compact." And such an idea is certainly not confined to ancient writers and philosophers: to support similar ideas, we have, for example, Sigmund Freud, William James, and many others. The only ones to disagree would seem to be the exponents of humanistic psychology, who have tried to connect creativity and sanity, namely, Abraham Maslow, Carl Rogers, and Rollo May.

And the most famous case of 'mad genius' is undoubtedly that of Vincent van Gogh, who apparently suffered from various psychopathological symptoms, famously ending up cutting off his own ear and subsequently shooting himself in the chest, dying after about thirty hours[6]—although there are those who felt it was not really suicide.[7] But van Gogh's is certainly not the only case: we need only think of the depressive attacks of the British writer Virginia Woolf, who also died by suicide, or the American poet Sylvia Plath, also the protagonist of several suicide attempts, right up to the final one. Or again—still in the field of poetry—think of Anne Sexton's suicide. In the light of such famous cases, a question arises: is there a price to pay for being a certified genius? It must certainly be said that the relationship between mental illness and suicide is very complex, and there is certainly a cultural factor to take into account. In fact, let us not forget that not only are psychopathology and voluntary death not necessarily connected, but that, in certain cultures, the latter is considered a more than honourable way out: just think of the Japan of the Samurai, or Ancient Rome, in which suicide was indeed contemplated and accepted, in a certain context, linked to honour.

Not only that, but mental illness does not necessarily have to result in suicide: think of John Nash, mathematical genius and schizophrenic, who died together with his wife in a car accident—certainly not a voluntary circumstance—and who had found a way to live with his condition.

Not only depression and schizophrenia, but also bipolar disorder, anxiety and panic attacks: think, for example, of Ludwig Bolzmann, who committed suicide in Duino, near Trieste. In addition, there is a whole range of other problems, such as alcoholism and drug abuse. For example, Henri de Toulouse-Lautrec, whose drinking problems went so far as to push him to fill his walking stick with liquor, so that he would always have his dose of alcohol

---

[6] According to his brother Theo, his last words were "The sadness will last forever.".

[7] Naifeh, Steven & White Smith, Gregory, *Van Gogh: The Life*, Random House, New York 2011.

at hand—a circumstance that ended up killing him, along with syphilis, at only 36 years of age.

All of these are just anecdotes, however, and it is time to take up the question of the relationship between genius and mental illness in a more thorough and systematic manner. Let us ask ourselves, then: is mental illness a necessary side-effect of genius? Perhaps to be a genius, one's mental health must be less than perfect? Perhaps pathologies of the mind even represent an advantage? Let's break this question about genius and mental health down into more specific questions—courtesy of Simonton.[8] First: is there any difference, in terms of the frequency of mental pathologies, between geniuses and non-geniuses? Second: is there any link between mental health, illness and the specific field of application of genius? That is, is there a difference related to the specific field in which a certain person manifests his or her genius? Thirdly, is there any degree of difference running parallel between genius and mental illness? That is, do people with a greater degree of genius also tend to manifest greater psychopathological conditions, which decrease as artistic abilities decrease?

These are obviously still open questions, but ones that various scholars have tried to answer over time. In particular, we cite two pieces of research that are of particular interest to us, namely, that of the American psychiatrist Arnold Ludwig[9,10] and that of his British colleague Felix Post.[11] Let us begin with Ludwig, who has carefully evaluated the medical histories of about a thousand famous people—both in the creative and leadership spheres. Just to understand the calibre of the characters studied by the psychiatrist, we cite, in alphabetical order—among writers and poets—Guillaume Apollinaire, Simone de Beauvoir, Berthold Brecht, Andre Breton, Albert Camus, Truman Capote, Anton Chekhov, Agatha Christie, Jean Cocteau, Joseph Conrad, Gabriele D'Annunzio, Arthur Conan Doyle, William Faulkner, Anatole France, Robert Frost, Federico Garcia Lorca, Ernest Hemingway, Hermann Hesse, Aldous Huxley, Henrik Ibsen, Henry James, James Joyce, Franz Kafka, Rudyard Kipling, D. H. Lawrence, C. S. Lewis, André Malraux, Thomas Mann, Somerset Maugham, Vladimir Nabokov, George Orwell, Boris Pasternak, Ezra Pound, Marcel Proust, Rainer Maria Rilke, George

[8] Simonton, Dean Keith, *The Genius Checklist*, The MIT Press, Cambridge 2018, p. 22.

[9] Ludwig, Arnold M., *The Price of Greatness: Resolving the Creativity and Madness Controversy*, Guilford Press, New York 1995.

[10] Ludwig, Arnold M., *Method and madness in the arts and sciences*. Creativity Research Journal, 11, 93–101, 1998.

[11] Post, Felix, *Creativity and psychopathology: A study of 291 worldfamous men*, British Journal of Psychiatry, 165, 22–34, 1994.

Bernard Shaw, Johan August Strindberg, Dylan Thomas, Leo Tolstoy, Mark Twain, H. G. Wells, Oscar Wilde, Tennessee Williams, Thomas Wolfe, and William Butler Yeats.

Among the artists are Paul Cézanne, Edgar Degas, Marcel Duchamp, Paul Gaugin, Edward Hopper, Gustav Klimt, Oskar Kokoschka, Le Corbusier, Henri Matisse, Edward Munch, Georgia O'Keeffe, Pablo Picasso, Jackson Pollock, Pierre August Renoir, Auguste Rodin, Henri Toulouse-Lautrec, and Andy Warhol. Composers include Louis Armstrong, Bela Bartok, Claude Debussy, Antonín Dvořák, Duke Ellington, George Gershwin, John Lennon, Gustav Mahler, Charlie Parker, Sergei Prokofiev, Giacomo Puccini, Sergei Rachmaninoff, Maurice Ravel, Arnold Schoenberg, Dmitri Shostakovich, Richard Strauss, Igor Stravinsky, and Giuseppe Verdi.

And then the scientists: Alexander Graham Bell, Niels Bohr, Marie Curie, Thomas Alva Edison, Albert Einstein, Alexander Fleming, Henry Ford, Ernest Orlando Lawrence, J. Robert Oppenheimer, Nikola Tesla, Alan Turing, and Orville and Wilbur Wright. And finally, the philosophers: John Dewey, Friedrich Nietzsche, Josiah Royce, Bertrand Russell, George Santayana, Jean-Paul Sartre, and Alfred North Whitehead. A very large sampling, as you can see, enough to get an idea, at a glance.

For his part, Felix Post examines far fewer characters, and also focuses on the nineteenth century. In particular, he analyses 291 of them, including Darwin, Schopenhauer, Melville and Listz. In any case, Ludwig's and Post's works overlap to some extent, and the results are also similar. At this point, there are those who might question the permissibility of making post-mortem psychiatric diagnoses. However, if we have done well with Cox's work, we will also do well with that of the two psychiatrists—including because, by now, this type of 'remote' research is not uncommon, indeed, it is the fruit of a well-established methodology. So, let us see how the creative geniuses evoked by Ludwig and Post stand in relation to the questions raised by Simonton. To begin with, Post places in the group of serious cases characters clearly affected by mental disorders and obvious problems: Hemingway, alcoholic, paranoid, and suicidal; Rachmaninoff, suffering from a severe form of depression; the father of sociology, Auguste Comte, who was admitted to a psychiatric hospital and attempted suicide; Bolzmann, whom we have already mentioned; and Edvard Munch, who suffered from an extreme form of anxiety—as well as being paranoid—which also inspired his very famous painting *The Scream*. Given these cases, it seems possible to exclude the idea that artistic excellence is accompanied by perfect mental health; not only that, but art seems to be a very poor form of self-therapy, if we consider the unresolved psychiatric problems of the aforementioned characters—or those resolved in the worst of ways, as is the case with Woolf and Hemingway.

Let us move on to the second question, that of the relationship among genius, madness and the field of application. In particular, let us try to distinguish between scientific geniuses and artistic geniuses. According to Ludwig, in the case of the natural sciences, the presence of mental disorders throughout the life span of the relevant geniuses is around 28%, a low percentage, if we calculate that composers reach 60%, figurative artists 73%, fiction writers 77%, and poets as high as 87%. Post gives us similar percentages, with scientists seeming to fare better—only 18 per cent have serious pathological conditions.

Let us now look at the other side of the issue, namely, the percentages of those who can boast of iron mental health. According to Post, 31 per cent of the scientists he included in the study showed no evidence of any kind of pathological condition, not even a mild one, while composers drop to 17 per cent, visual artists to 15 per cent, and writers to 2 per cent. It must be said that this meagre 2 per cent consists of only one case, Guy de Maupassant, who was mentally ill and had an attempted suicide behind him, but owed this to an organic cause, syphilis—which is why Post decides to exclude him from the list of the mentally ill. In short, it appears from the two studies combined that mental problems mainly affect practitioners of the visual arts, writers and composers, and not scientists. Furthermore, according to Ludwig's data, architects and social scientists seem to be in the middle of the road between mental illness and sanity.

But the situation is, according to Simonton, much more complicated than it seems, and to clarify it, we must use the concept of 'paradigmatic science'; introduced for the first time by the philosopher of science Thomas Kuhn in the classic *The Structure of Scientific Revolutions*,[12] it refers to the fact that science would proceed precisely by paradigms, i.e., by organic structural visions, which succeed one another, replacing one another. In essence, Einsteinian physics—and quantum mechanics—has replaced Newtonian physics, just as the latter has imposed itself on the previous paradigm, and so on. Scientific research would thus consist of two types, the 'normal' type, embedded within a certain paradigm—to which it contributes, perfecting it—and the revolutionary type, which instead destroys the paradigm in question, replacing it with a new vision of things. Just like research scientists—and artists—of genius would, for Simonton, be classifiable into "normal" and "revolutionary." Not only that, but the psychopathological risk would be linked precisely to the role the genius in question plays within a certain paradigm: in other words, a 'normal' genius would run a lower psychic

[12] Kuhn, Thomas, S., *The Structure of Scientific Revolutions* (2nd ed.), University of Chicago Press, Chicago 1970.

risk than a 'revolutionary' genius. The discourse changes somewhat if we consider the social sciences—sociology, anthropology, and so on—which are in a pre-paradigmatic situation, i.e., they would not form unitary blocks, but would be divided—as the philosopher of science Imre Lakatos used to say—into different research programmes fighting against each other—a fact that, for Simonton, would also precisely explain the aforementioned data concerning the mental health of social scientists. Is there empirical evidence to support Simonton's thesis? It would seem so. In particular, we cite a study by two Korean researchers, Young-gun Ko and Jin-young Kim, and based on the analysis of 76 scientific geniuses.[13] The scholars divided the eminent people in question into four groups according to the severity of their mental pathology. Thus, 22 had no pathology, 13 had mood disorders, 27 had personality disorders, and 14 were schizophrenics—including psychotics. The geniuses in question were also evaluated according to their role within their scientific discipline, being divided into two groups, namely, "proponents of a certain paradigm" and "destroyers of a certain paradigm." And, according to Ko and Kim, sane scientists tended to be included in the group of "advocates," while the mentally disturbed were more likely to end up in the group of "destroyers." And, if we want an excellent example of this tendency, we can certainly cite Isaac Newton, who—apart from being a physics revolutionary—most likely suffered from paranoid schizophrenia and other disorders. This can be deduced from, among other things, his correspondence—for instance, the scientist wrote a decidedly paranoid, dark letter to John Locke, accusing him of setting a trap for him through women. Even before Ko and Kim, Ludwig himself put forward the hypothesis that this relationship between normality and revolution—both in the sciences and the arts—reveals an underlying relationship with mental illness. In practice, for the American psychiatrist, normal science requires logic and objectivity, and thus emotional stability, in those who practice it, while revolutionary science would rely more on intuition, and thus emotionality and instability. Not only that, but this association would also be noted in the arts: more specifically, the arts that are more structured from a formal point of view—e.g., architecture or composition—would be less affected by mental pathologies than the arts based on performance—acting, dancing, singing, and so on. In the literary field, then, poets would fare worse than novelists, and novelists would fare worse than non-fiction writers—journalists, essayists, and so on. Artists embodying more emotive styles—the expressionists, for example— would fare worse than the realists, surrealists, Dadaists, and all those more or

---

[13] Ko, Y., & Kim, J., *Scientific geniuses' psychopathology as a moderator in the relation between creative contribution types and eminence*, Creativity Research Journal, 20, 251–261, 2008.

less symbolic styles; the latter, in turn, would be outperformed in health by practitioners of more formal styles, such as abstractionism, futurism, cubism and conceptual art. Thus: both the sciences and the arts would be subject to disciplinary variations relative to the degree of mental pathology of its practitioners, and the latter would be related to the style required by each discipline, whether emotional-intuitive or rational-objective.

And so, we come to Simonton's third question on genius, namely, that concerning the gradualness of the disorder in relation to the degree of creative eminence. Let us start with this last concept, and immediately note that eminence is a variable that has a very wide range; for example, in the case of English poetry, we see authors who are highly cited and studied, such as Shakespeare, and authors who, although esteemed, have no monograph dedicated to them. And so, here, we are faced with a fine and good paradox: the most eminent creator, the most studied, to the point of quantitatively eclipsing the vast majority of others—in our case, Shakespeare—is also the least representative. Shakespeare, Chaucer and Milton make up 33 per cent of critical studies, but represent only one per cent of authors (according to Simonton's data).[14] And this is why general rules for the disciplines that we are considering cannot be deduced from exclusive analysis of the most eminent characters. It is precisely for this reason that the mental conditions of the latter cannot be regarded as the rule of the respective professions. It follows that research relating to so-called 'normal' or 'everyday' creativity cannot be associated with research relating to 'eminent' creativity. Thus, it would be wise for us to focus precisely on the most eminent creative geniuses, leaving out all the minor characters and keeping well in mind that there are at least two types of eminence, so the correlation between pathology and genius is not obvious. Let us therefore take Ludwig's work in hand; the scholar has, in fact, applied to his subjects a rating scale that he developed and christened CAS, which stands for Creative Achievement Scale.[15] For each author, we ask ourselves, with regard to his work, whether it has exerted a lasting impact, whether it has wide application—i.e., whether it has influenced the entirety of western civilisation, and so on—and above all, whether and how original it is. In relation to the author, we also ask whether he overcame the limitations of his own era, whether he influenced contemporaries and followers, and so on—there are a total of eleven items in the CAS. Subsequently, Ludwig's analysis precisely indicates a strong correlation between the CAS scores and psychopathological disorders, i.e., that a high degree of genius

---

[14] Simonton, Dean K., Op. Cit., p. 40.
[15] Ludwig, Arnold M., *The creative achievement scale*, Creativity Research Journal, 5:2, 109–119, 1992.

would indeed be accompanied by this or that pathological condition, but up to a certain degree—high, assuredly, but with certain limits—beyond which creativity obviously becomes impossible—an excess of mental suffering can, in fact, only serve to block any creative effort.

At this point, we take Post's work, and note that, according to his data, a high level of eminence is linked precisely to a high degree of psychopathology, which clearly separates the geniuses from the non-geniuses. Only in the field of the arts, however: according to Post, in the sciences, such a relationship is not noticeable. And when, in the sciences, we note the presence of mental illness, it is usually of moderate intensity. So, combining everything we have seen so far, we can conclude—as far as Simonton's third question is concerned—that eminence is indeed related in a special way to mental illness, in a way that separates it from non-eminent or less eminent creativity. The answer is very complex, however: from the data examined, there is, in fact, a complex association between eminence and mental illness, in the sense that mental illness often accompanies genius, but it does so in an individualised way, and there are often exceptions: if we examine our case histories individually, we will note that there are several geniuses who do not suffer from mental illness, others who do but manage to find moments of sanity that are useful for creating, and still others who even use their disorders. Let's just take the example of the Austrian composer Anton Bruckner, who suffered from a strong obsessive–compulsive disorder, which, however, allowed him to obsessively revise his compositions to perfection. Or think of another trait related to creativity, namely, cognitive disinhibition—it sounds good, but it is a problematic condition, as it consists of the inability to selectively filter out what comes from the outside world to capture our attention. This is an advantage from the point of view of creativity—it allows us to grasp aspects and details that others do not notice—but it is also linked to an increased risk of mental illness, particularly schizophrenia, whose carriers are overwhelmed by the stimuli they receive. As far as historical examples are concerned, we need only mention Alexander Fleming, who noticed something that would have escaped the average scientist, namely, the mould infesting his bacterial cultures, an observation that led directly to the discovery of penicillin. Too bad that, as mentioned above, cognitive disinhibition can also lead to loss of contact with reality, hallucinations and whatnot, as in the aforementioned case of John Nash. At this point, since we have realised that genius is often—but not always!—accompanied by mental illness, we wonder how the victims in question manage such disorders. Some help comes, according to Simonton, from fairly high general intelligence, from the IQ in practice, which therefore comes into play: in fact, it is precisely general intelligence that would help the genius to discriminate between the stimuli offered to them by cognitive disinhibition and to moderate or manage the psychiatric symptoms that afflict them. In short, the only thing separating madness and genius, even if only weakly, would be a thin wedge of intelligence. A theory, let us say, that convinces us up to a certain point, given the exceptions, including some excellent ones, such as John Nash, whose IQ and mental disorders are not in doubt. And in any case, all of this talk of genius, madness and sanity leaves us with a bitter taste in our mouths: we were hoping for a clearer answer, and not all of the nuances we got instead.

At this point, let us confess what we already knew from the start, namely, that, although the data we have set out shows a clear correlation between psychopathology and genius, the true nature of this relationship has yet to be clarified. The possible causal explanations are numerous. Could it be, for example, that the achievement of the summit—in the sense of scientific or artistic eminence—is so stressful that it endangers the psychic balance of the person pursuing this goal? In this case, madness would only be a consequence of genius, and the causal direction would go from the latter to the former. Someone who does *not* think so is the German-British psychologist Hans Eysenck, for whom the link between psychosis and creativity is not accidental.[16] On the contrary, the scholar has elaborated a Darwinian theory directly inspired by Campbell and Simonton's model, according to which the creator, in order to be able to create, needs a vast reserve of possibly remote associations—far removed from common sense—that will enable them to produce in an original, i.e., divergent, way. Eysenck's meta-analysis reveals a solid relationship between psychoticism and the ability to produce unusual ideas in large quantities. People with such characteristics would also be characterised by particular ways of thinking, i.e., of examining and analysing the world around them, according to 'overinclusive,' psychological mechanisms that reduce the boundaries between ideas, thus leading to – generalised (overgeneralised?) concepts. According to Eysenck, people who possess the right level of psychoticism have a tendency to receive seemingly irrelevant ideas into their minds in a random and uncontrolled manner. Although mostly composed of effectively useless and 'overinclusive' ideas, this flow would favour the emergence of interesting or useful mental associations. Another hypothesis is that the lack of conformity often associated with psychoticism would allow the creative genius to impose fewer constraints on the flow of thought, an essential ingredient of creativity. In fact, the great innovators in

---

[16] Eysenck, Hans J., *Genius: The natural history of creativity*, Cambridge University Press, Cambridge 1995.

Western history have distinguished themselves precisely according to a non-conformism that we would almost describe as pathological: it takes this kind of attitude, and with great intensity, if one wants to rage against established tradition. One only has to think, for example, of the extremely tough battle Ignaz Semmelweiss waged for many years against the medical establishment to convince his colleagues that puerperal fevers were contagious.

5.3   First-Born Versus Last-Born

Here is a lighter question: when should one be born, within a certain family, to become a certified genius? First, last, or in between? This question—not directly related to genius, however, but to one's role in society—was originally asked by an American psychologist, Frank Sulloway, according to whom birth order exerts a significant influence. His thesis—set out in his book *Born to Rebel*[17]—claims that this order exerts a very strong influence on individual psychological traits, the famous "Big Five." During the 1980s and 1990s, trait theory—a field of personality psychology—suggested that individual psychological traits could be classified into five groups. This taxonomy thus provides five headings: openness to experience (curiosity vs. caution); conscientiousness (efficiency vs. lack of care); extraversion (outgoingness vs. introversion); agreeableness (empathy vs. rationality/coolness); and neuroticism (sensitivity vs. resilience). All of these groups are combined in the famous acronym OCEAN. Well, for Sulloway, first-borns would be more conscientious, more leadership-oriented, less agreeable, and less open to the new than last-borns. Now, this theory has been much criticised—as is often the case when dealing with psychological constructs—but we will try to see if there might be something in it that is relevant to our studies of genius. Let us begin, using a specific pair of siblings, with a fundamental distinction, namely, that between shared and unshared environment. The former concerns the environment they have in common, namely, the family environment, including the neighbourhood. Let us take the example of the famous brothers James, William and Henry, Jr. Born about a year apart, the two could certainly be said to share the family environment—dad, mum, neighbourhood, family outings. The unshared environment is precisely what differentiates the two: e.g., any paternal or maternal favouritism, or the ways in which siblings interact with each other and with other siblings—William and Henry had two additional brothers and a sister. As time passes, the unshared environment takes on a

---

[17] Sulloway, Frank J., *Born to Rebel: Birth Order, Family Dynamics, and Creative Lives*, Vintage, New York 1997.

greater role: as the siblings grow up and go to school, they enter different classes, have different teachers, make different friends, and so on. The moral of the story: the shared environment—difficult, by the way, to distinguish from genetics, precisely because it is shared—plays a preponderant role, especially in the early years, those that should, in theory, give the initial set-up to the subjects we are considering, whether they are brilliant or not. Now, as is well known, William became a celebrated philosopher and psychologist, Henry a well-known writer. What is the role of the shared environment? It should be preponderant, since they were all born within a short distance of each other—five children in seven years. Why did they not all become psychologists, or philosophers, or writers? Does this difference perhaps have to do with birth order? This question is, as we can imagine, still without a definitive answer, but the study of genius can give us some indication. So, let us take a closer look at this idea, and call in our usual source, Galton, and, in particular, his 1874 book *English Men of Science*. Based on questionnaires submitted to 99 scientists, the book points out that, in 22 cases, they were only sons, in 26 cases, firstborns, and in 15 cases, lastborns. In addition to first-born and last-born, we have 13 from the oldest 'section' of the offspring, 12 from the youngest, and 11 exactly in between. Thus: according to Galton, primogeniture seems to have some effect on destiny towards individual genius—at least as far as the sciences are concerned. However, we note a significant lack in these calculations, namely, the presence of sisters. And primogeniture also seems to gain support from Ann Roe's later study of 64 genius scientists, of whom as many as 39 were first-born—actually, 15 of them were only children. 39 represented 61 per cent of the total.[18] In contrast, it was psychologist Ravenna Helson who looked at women, particularly eminent mathematicians and graduates of Mills College—a women's institution. Result: with few exceptions, the intellectually successful women had no siblings. Moreover, a strong identification with the father, who generally tended to be a professional, was generally present in them.[19] Mind you, however: these are trends, not rigid rules, so much so that there are excellent exceptions, such as Charles Darwin, the penultimate son—fifth after three brothers and a sister, and followed by a further sister. Not to mention Galton himself! Last-born, after four sisters and two brothers, none of whom managed to reach his level of eminence.

---

[18] Roe, Anne, *A psychologist examines 64 eminent scientists*, Scientific American, 187(5), pp. 21–25, 1952.

[19] See: Helson, Ravenna, *Women mathematicians and the creative personality*. Journal of Consulting and Clinical Psychology, 36, 210–220, 1971. Also: Helson, Ravenna, *The creative woman mathematician*, in: L. H. Fox, L. Brody, & D. Tobin (Eds.), *Women and the Mathematical Mystique*, Johns Hopkins University Press, Baltimore 1980, pp. 23–54.

Another question, related, although not identical, to the previous one on birth order: does family size count? If, in fact, someone is an only child, there is no real birth order, and indeed perhaps said individual should be ascribed to a separate category. More generally, is there a relationship between eminence and family size? It would seem so, according to the research of a US psychologist, Robert Zajonc, who developed the so-called 'confluence model' in the late 1970s. Together with his colleague Greg Markus, the scholar developed a mathematical model associating IQ and family size. In practice, according to this model, the family environment would influence the intelligence level of the child, but not homogeneously. In fact, the first child would find itself in an environment of adults only, with relatively high intellectual stimulation, while the subsequent children would progressively have to deal with more and more infants, thus experiencing less stimulation, and this would cause a reduction in parental stimulation in relation to IQ—in practice, this would lead to an IQ lower than that of the first-born. However, this would be a very small effect of about 3 IQ points. IQ that, in any case, only relates to eminence to a certain extent, as we have already seen.

The argument that Simonton makes is obviously much more articulate than we have rendered it, and refers to the possibility that, while scientists would tend to be first-born, artists and writers—and even revolutionaries— would tend to be last-born.[20] The scholar emphasises, however, that, data in hand, the effects of birth order, while existent, are unimportant.

5.4   Hard Work and Concentration or a Carefree and Varied Life?

As for the lifestyle of geniuses, what does scientific research tell us? While there is a tendency to think that eminent people become eminent precisely because of their obsessive commitment, there are also those who point out that creativity requires freedom—including the freedom to move and travel— carefreeness and interests of many different kinds. Let us therefore take a look at this theme. And let us begin by quoting Copernicus, in particular, his revolutionary work, the *De revolutionibus orbium coelestium,* which, in fact—pardon the multiple repetitions—revolutionised astronomy and the very concept of the relationship between humanity and the cosmos. Why him? Because, according to the story, he received the first printed copy of his

---

[20] See Simonton, Dean K., Op. Cit. pp. 89–101.

work on his deathbed. True or not, it is a fact that his book was published the year he died. But what is actually important is that Copernicus worked on his book for a good part of his life: twenty years of basic studies—mathematics and astronomy—and another thirty of speculation, in practice, a good half century of dedication. Here's the point: it seems that, to rise to the level of genius, dedication and hard work are essential, according to the meagre biographical notes that we have provided. Of course, he did not literally work on the book for fifty years, but certainly put in at least ten thousand hours, according to a famous formula—that of deliberate practice. The idea of deliberate practice—that is, the procedure that leads to excellence—was originally launched by Herbert Simon, the American psychologist, who later won the Nobel Prize for economics, and it was his follower, Anders Ericsson, who took it up again in a systematic way. We will return to this idea in detail later, but, for now, we anticipate an observation on the double soul of psychology, given to us by a psychologist who was also one of the Termites, Lee Cronbach. The scholar, in fact, published an important article in 1957, entitled *Two Disciplines of Psychology*. In it, Cronbach emphasised how psychology actually consists of two separate disciplines, experimental psychology and correlational psychology. While experimental psychology—which goes back to Wilhelm Wundt, and of which Simon and Ericsson are adherents—starts from the assumption that the subjects of the experiments are substantially all the same, that is, all endowed with the same universally valid characteristics, correlational psychology—to which Galton and Simonton subscribe—insists on individual differences, and tries precisely to discriminate subtly between them, relating causes and effects. Why are we interested in this distinction? Because of the fact that experimental psychologists like Ericsson would tend to deny the existence of individual talent—if you fail to achieve a certain result, it is your fault, basically—while correlational psychologists would hold an opposite view of things. In particular, Simonton defines talent as the *innate*—let us emphasise *innate*—ability to accelerate the learning of a certain skill. Be that as it may, the scholar also brings us a series of concrete examples according to which the central rule of deliberate practice—ten thousand hours of intensive practice—would be violated. Let us cite just one, striking case, that of Felix Mendelssohn, who began taking piano lessons at the age of six, and began studying composition at ten. Between the ages of twelve and fourteen, he composed numerous symphonies that are still broadcast on the radio today; thus, it only took him six years of musical practice and two years of compositional study to qualify as a composer, far less than the ten years/ten thousand hours required by Simon and Ericsson. Of course, it could be argued that Mendelssohn's first real masterpiece is his String Octet in E-flat Major, composed at the age of sixteen, which still makes an impression, given that it only required six years of study of compositional techniques.

And what about those who, instead of devoting themselves with rigour and dedication to a single discipline, jump from one field to another? Is this acceptable for a genius? Is it OK to be a polymath? It would seem so. In fact, it could even be helpful. Let us consider the case of Galileo Galilei. The father of contemporary science always showed a strong interest in both the visual arts and literature—and the latter probably helped him to make his works examples of polemical and persuasive texts. Galileo's artistic training included chiaroscuro, which enabled him to train his eye not only to see the lunar mountains through the telescope, but also to draw them. Before going any further, however, let us take a look at the concept of the polymath itself, and see how it is intertwined with that of the genius. By this term, we mean a person who demonstrates a thorough—hence, not amateurish—interest in more than one field of knowledge, and possibly in many. Thus, at first glance, Leonardo and Leibniz certainly come to mind, but it must be borne in mind that polymathy was a widespread condition, when the knowledge possessed by humanity was limited and the struggle for survival was a daily necessity: the ancient peasants or hunter-gatherers of times gone by were, in fact, experts, in many different arts, and had to be if they wanted to survive. The cultural and technological evolution of society led to specialisation, but also nurtured the ambition of some human beings to possess all the knowledge available at the time. The current concept of polymathy is thus connected to the partitioning of knowledge—and academic knowledge in particular— into circumscribed disciplines. Some have even tried to classify polymaths into different types: we have passive polymaths, who produce nothing original, and active polymaths, who contribute to the culture; general polymaths, who deal with disciplines far removed from each other, and limited polymaths, who concentrate on related disciplines; simultaneous polymaths and serial polymaths; and then there are systematizers, such as Auguste Comte or Francis Bacon, who try to order human knowledge into a single system. Offering this classification is a British historian, Peter Burke, in his excellent work *The Polymath: A Cultural History from Leonardo da Vinci to Susan Sontag*.[21] Among the polymaths, in addition to Leonardo and Leibniz, Burke mentions Johann Wolfgang von Goethe, George Eliot, Aldous Huxley, and Jorge Luis Borges, as well as Vladimir Nabokov, who, besides being the author of *Lolita*, was also an entomologist, chess player and literary critic. But we

---

[21] Burke, Peter, *The Polymath: A Cultural History from Leonardo da Vinci to Susan Sontag*, Yale University Press, New Haven 2020.

could mention many others. What is important, however, is another matter, namely, the fact that the relationship in cases like these between expertise in many different fields and the ten-thousand-hour rule is not very clear: does each field require ten thousand hours? So, a polymath with expertise in, say, four fields must have committed to a total of forty thousand hours? The answer Simonton gives refers to the notion of talent as an innate ability to acquire a skill or knowledge with exceptional speed, and to an above average degree. This would therefore explain a phenom such as Johann Wolfgang von Goethe, who wrote novels of the highest order, poetry—both lyrical and epic—and outstanding contributions to anatomy and botany. Not to mention his highly influential colour theory. And all this while working as a civil servant and diplomat. Ten thousand hours hardly seems to explain Goethe. While we're at it, let's take a look at the daily lives of the creative greats. Mason Currey, in his golden booklet *Daily Rituals. How Artists Work*,[22] poses some essential questions, which, while relating mainly to writers, apply, by extension, to geniuses of all kinds: how do they find the time to devote to their art or interests? According to Currey, the secret lies in daily rituals, i.e., in the fact that many of the greats manage to apply their creativity to the management of everyday life as well. In a very subjective way, though: everyone has their own highly individual creative process. Pick up the book, read it, and see if any of the rituals used by the mentioned authors suit you. Remember, however, that, underlying each of them, despite the sometimes chaotic lives of the characters in question, there is a great deal of discipline.

But then, does studying for ten years/ten thousand hours really help? It depends: just invent an entirely new field, and you don't have to acquire any previous expertise. Think of the seventeenth-century Dutch scholar Antonie van Leeuwenhoek, who took a microscope and closely observed, for the first time, microscopic animals that eluded the human eye and were not contemplated in Aristotle's works: no ten thousand hours of deliberate practice for him.

5.5 Pursuing Perfection or Shooting at Random?

What is the relationship between genius and perfectionism? It would seem obvious that a genius is also a perfectionist, given that he or she must create a solid, lasting and inspiring work. Among perfectionists, Stanley Kubrick

---

[22] Mason Currey, *Daily Rituals, How Artists Work*, Knopf, New York 2013.

and Steve Jobs immediately come to mind, while, in the opposite camp, we are reminded of Alexander Fleming, whose less than perfect laboratory cleaning practices earned him the discovery of penicillin. Also among the notoriously perfectionist geniuses are Isaac Newton—he requested the removal of his name from two of his publications because of some errors— and Michelangelo—whose perfectionism slowed down his work considerably. The problem lies—according to Simonton's correlational analysis—in the fact that, while the work of genius seems to be somehow connected to the idea of perfection, psychological analysis does not seem to find traces of the psychological traits connected to the perfectionist personality in the average profile of geniuses. Put another way, geniuses are not perfectionists, except when it comes to their own work. On the contrary, imperfection seems to be a necessary characteristic, at least according to Simonton, who insists, as always, on the 'trial-and-error' aspect that he sees as typical of genius production. We must agree with him on one thing, however: there is no single creative process that infallibly leads to the production of a masterpiece. If it existed, we would probably have already found it, and we would also know how to apply it to produce works of genius automatically. Instead, in reality, creative geniuses—both scientific and artistic—use a vast and varied amount of procedures, choosing them according to their personal inclinations and idiosyncrasies. Simonton provides us with a partial list of these, from the most obvious to the most abstruse: we mention, if nothing else, daydreaming, de-focussed attention, insight dear to gestaltists, analogy, intuition, asking the right question, actual dreams, reframing, the 'trial-and-error' dear to Simonton himself, the Janusian and homospatial processes, the Geneplore[23]—a term derived from the fusion of 'generate' and 'explore'— the preparation-incubation-illumination-verification process, brainstorming, and, finally, the simple game. In short, the perfection of the work stands out against a panoply of methods that, far from being perfect, sometimes work and sometimes don't, methods from which to choose according to one's inclinations. So, here is the recipe: explore the tools of creativity, choose what suits you, and start producing as much as you can, collecting both successes and failures. If Lady Luck is on your side, you will end up like Mozart—who more or less systematically produced perfect operas; if luck is against you, you will not produce anything significant, or perhaps you will end up like Johann Pachelbel—who?—the German composer who, despite

---

[23] In essence, it consists of a procedure based on the generation of forms that are then studied; in other words, one starts without a precise definition of the problem, exploring it as one develops solutions. See Finke Ronald A.; Ward Thomas B.; Smith, Steven M., *Creative cognition: Theory, research, and applications*, MIT Press, Cambridge 1992.

having composed more than five hundred works, is only remembered for his Canon—one of the most frequently played compositions during wedding ceremonies.

5.6   Child Prodigy Versus Late Bloomer

And here we come to a question that is only seemingly academic: in reality, when comparing child prodigies and late bloomers, what many readers—perhaps of a certain age—will actually want to know is whether there is still hope for them, that is, whether it is not too late to rise to the level of genius. Let us take a closer look at this question, then. Certainly, if you aspire to become a genius, the path of the child prodigy seems the most suitable: being able to master this or that discipline within or even before the age of ten would, in fact, allow you not only to break the ten years/ten thousand hours rule, but also to rest on your laurels during the following decades. This does not mean that, once you have acquired child prodigy status, you necessarily have to stop producing works of genius: in fact, there is no shortage of cases of child prodigies who have gone on to work. For example, among scientists, there is Svante Arrhenius—the father of physical chemistry and Sweden's first Nobel Prize winner—Enrico Fermi, Sigmund Freud, Francis Galton, Carl Friedrich Gauss, Ted Kaczynski—yes, the unabomber—John von Neumann, Blaise Pascal, Jean Piaget and Norbert Wiener; among writers and poets, Samuel Johnson, Pablo Neruda, Alexander Pushkin, and Arthur Rimbaud; among artists, Basquiat, Gian Lorenzo Bernini, Albrecht Dürer, Pablo Picasso; among composers, Béla Bartók, Georges Bizet, Frédéric Chopin, Felix Mendelssohn, Wolfgang Amadeus Mozart, Sergei Prokofiev. And finally, let us mention Blaise Pascal, who managed to accumulate within himself the certificate of genius, child prodigy and polymath: among other things, as a teenager, he began to speculate about the possibility of developing machines capable of performing calculations, not bad for a young man of that age. Unfortunately, however, child prodigies are not always destined to become geniuses, both because not all areas of creativity are open to child prodigies and because it often happens that the child's technical development is not accompanied by social development, with all of the consequences that this entails. If we then go on to classify the areas in which child prodigies tend to manifest their exceptional performances on average, we will note that there are three of them: chess, musical composition and mathematics. Which prompts us to ask: what do these three disciplines have in common? Simple: all three are decidedly abstract fields based on precise rules and clear objectives. Yes, even musical composition—think orchestration, counterpoint, rhythm and harmony. In short, these three fields have procedures that allow those who work hard and are guided properly to move from novice to expert level relatively quickly. The visual arts are an entirely different story, and the same applies to literature, disciplines that often require extensive life experiences to produce significant results.

Not that child prodigies are really precocious in everything; in fact, it often happens that they are deprived of adequate socialisation, and pay the price later. This is the case, for example, with William James Sidis, an American mathematician and child prodigy, literally trained by his father—the psychiatrist Boris Sidis—in order to become a genius—note the name given to him, William James. Sidis got off to a good start, apparently; he was admitted to Harvard College at the age of eleven and made a presentation to the Harvard Mathematics Club on four-dimensional bodies. Once he left his father's supervision, Sidis lost interest in mathematics, became involved in politics without much success and died in obscurity at the age of 46. Let us also mention the case of Ted Kaczynski, a child prodigy in the field of mathematics, who entered Harvard at the age of sixteen and was subjected to a series of traumatic psychological experiments by Henry Murray—and perhaps it was precisely this experience that dealt the final blow to the psyche of the future Unabomber.

If, however, you think that the path of the child prodigy is not suitable for you—perhaps, for you, as for yours truly, it is too late—then it would be worth considering the path of the late bloomers, i.e., people who start very late in acquiring the fundamentals of a certain discipline, or who perhaps start early but take much longer than average. At this point, we want to understand why some geniuses 'wake up' late, and to do so, we first introduce the concept of the 'Crystallizing Experience.' By this term, we mean those existential events—in practice, those instances in life—due to which a certain person comes into contact with a talent, a concept, a practice or a master that somehow makes them discover an important creative direction, or causes them to change the one they had already taken. Let's think about it for a moment: not everyone is lucky enough to know from an early age what their vocation is, in fact, most people move somewhat awkwardly, going from one field to another in search of a direction. And then, there are those—often geniuses—who have a precise Crystalizing Experience, an epiphany if you like—although here, we mean this concept in purely psychological, not mystical, terms.[24] Think, for instance, of the great jazz musician

---

[24] Walters, Joseph; Gardner, Howard, *The Crystallizing Experience: Discovering an Intellectual Gift*, Harvard University, Cambridge 1984.

Herbie Hancock—the one from *Cantaloupe Island*, remember that one?—who began as a child prodigy with classical music, then switched to Jazz in his twenties. It was his encounter with the recordings of the vocal quartet The Hi-Lo that got him interested in Jazz, and witnessing a performance by Jazz pianist Chris Anderson finally convinced him of the need to receive formal training in that musical genre. An experience that precisely crystallised his talent, turning him into a genius. Or let us cite the Austrian composer Anton Bruckner, who, working mainly on religious music, had a Crystallising Experience at the age of 39 in the form of his instructor, who introduced him to the works of Wagner. In short, don't lose hope that a Crystalizing Experience could happen to you at any age.

Sometimes, it is life's adversities that make us discover a certain discipline. This is the case, for instance, of Anna Mary Robertson, also known as 'Grandma' Moses, a well-known American practitioner of folk art, particularly embroidery—admittedly, not a form of artistic expression that everyone would consider as 'genius.' Plagued by arthritis, she had to abandon this technique and began—at the age of 76—to take up painting, thus beginning an original artistic career very late indeed. So, if 'Grandma' Moses found her way at that age, we can all certainly follow her example. Or again, the phenomenon of late bloomers may be determined by a certain slowness in acquiring the fundamentals of a certain discipline, a slowness, in turn, due to the facts of life: not all grown-ups have the luxury of being able to devote themselves to their preferred interests full-time, but perhaps have to divide their time between work obligations and family commitments. Economic difficulties can play a role in all this: think of Marie Curie, who, in addition to gender discrimination, had to face several financial difficulties, which drew out her time in school, preventing her from graduating until she was 27 years of age.

But there is a further figure, which complements the child prodigies and late bloomers: that of the one-hit wonder. This term—taken from the field of music[25]—refers to those geniuses who have become famous for a single work. And this is hardly confined to pop music; such things happen in every field of knowledge and creation, that is, there are people who are only known for a single work, in spite of the fact that certified geniuses are usually known for many works. Simonton explains this easily, particularly with his theory of random variation and selection; in essence, the one-hit wonders would simply be unfortunate cases. In this connection, the American scholar cites

---

[25] If you were around in the eighties—and if you are a fan of the music of that era—you will probably remember Nena and her song *99 Luftballons*; also from the eighties, we mention *Tainted Love* by Soft Cell. In more recent times, we remember *Spaceman*, by Babylon Zoo.

the case—this time taken from mathematics—of Roger Apéry, known exclusively for a theorem named after him, which proves that a certain well-defined number cannot be expressed as a ratio of two integers—in mathematical terms, such a number is said to be irrational.

5.7   Dying Young or Living to a Ripe Old Age?

If you follow contemporary music folklore, you are likely to have heard of the famous—or infamous—Club 27, the list of famous musicians who died at the age of twenty-seven: this club—just a journalistic expression, really—brings together the likes of Jimi Hendrix, Janis Joplin, Jim Morrison, Brian Jones—the musician who launched the Rolling Stones—Kurt Cobain and Amy Winehouse. Here, Club 27 is the embodiment of the romantic myth of the genius dying young, possibly tragically. Notwithstanding the fact that I, and probably a number of my readers, are too old to die young, let us now explore the relationship between creativity and death—one that, besides being a nuisance, represents precisely the end of creativity, and the final seal, which establishes, from the outside, the meaning with which the life of this or that genius is imbued. In other words, let us take a look at the life cycle of genius.

In the case of the one-hit wonders, the story ends there, with that one work, upon which the reputation of the creative person in question depends. Let us think, for instance, of Pietro Mascagni, an Italian composer who, despite having composed numerous operas, remained famous for only two hits, *L'amico Fritz* and *Cavalleria Rusticana*. A brief sad story: Mascagni composed the two operas in question when he was 27–28 years old, became famous for them and continued composing until the age of 72, despite the fact that his career had, in terms of originality, long since ended. Mascagni himself declared that he was crowned before he was even king.

Exactly: is there a right time to become productive? To produce masterpieces? It would seem so, but that would be a fact related to the discipline we are considering. This was argued by an American psychologist, Harvey C. Lehman, in his book *Age and Achievement*.[26] And so, as far as prose and poetry are concerned, the best quality novels would be written, according to his research, between the ages of forty and forty-four, while sonnets, lyrics, elegies, and poetry in general would commonly be written between the ages of 25 and 29. These are, of course, only trends, admitting exceptions, but,

---

[26] Lehman, Harvey C., *Age and Achievement*, Princeton University Press, Princeton 1953.

in general, this shows that poets can be surprised by death even before the age of thirty and still manage to write memorable works. And indeed, there is no shortage of examples of poetic precocity linked to early death: John Keats, for instance, died at the age of twenty-five, while Percy Shelley—Mary's husband—died at 29, and Robert Burns at 37; in short, English Romantic poets seem to be united by a common destiny. As far as scientific research is concerned, there would be differences among the various disciplines. Lehman's reflections, corroborated by the later ones by Simonton,[27] point out that, in the case of eminent scientists, on average, the first major works appear around the age of thirty, and the major works around forty, but that major productivity would be maintained until the mid-fifties. An exception—in terms of precocity—are mathematicians, who begin to produce major works, on average, in their thirties, a bit like poets. And they even die prematurely, like the latter: Srinivasa Ramanujan died at the age of thirty-two, Niels Henrik Abel at twenty-six, and Évariste Galois at twenty, due to a duel. In essence, Galois, as a teenager, managed to solve a mathematical problem that had remained open for 350 years, related to the resolution of polynomials—let us not go into details—a very impressive achievement!

However, the question of scientific creation and the life cycle of scientists in particular will be more thoroughly elaborated in the next chapter. For now, let us limit ourselves to the question of the life cycle of great geniuses, and do so with the work of an American scholar who has dealt with this issue, David W. Galenson. An economist by training, Galenson has written an interesting book on the subject: *Old Masters and Young Geniuses*.[28] The scholar draws a distinction between two opposing creative styles related to the art world. The first style concerns conceptual creators, who rise to the level of genius early, thanks to sudden 'illuminations'. Galenson christens them "finders" or "young geniuses." Pablo Picasso would, in his view, represent this type of genius. Other geniuses work in an exploratory, gradual manner, over a long period of time, and, obviously, their major works would appear relatively late, after they had accumulated a great deal of experience. Galenson christens them "seekers" or "masters," and he uses Paul Cézanne as an example. Starting with painting, the American scholar then extends his model to the other arts—literature, poetry, cinema—aiming to include even scientific research. It's an interesting model, Galenson's; we don't know whether we entirely buy it it, but you certainly have more options now: you can choose whether to

[27] Simonton, Dean K., *Career landmarks in science: Individual differences and interdisciplinary contrasts*, Developmental Psychology, 27, 119–130, 1991.
[28] Galenson, David W., *Old masters and young geniuses. The Two Life Cycles of Artistic Creativity*, Princeton University Press, Princeton 2006.

wait for an insight, or to devote yourself more to craft. Be aware, however, that time is only on your side if you choose the second option. Choose your career wisely afterwards: if it is probably too late for you to become a renowned poet or mathematical genius, you can still opt for a different kind of profession and become a historian or philosopher. Kant, in fact, is a good example to remember: the great philosopher published the *Critique of Pure Reason* at fifty-seven, the *Critique of Practical Reason* at sixty-four and the *Critique of Judgement* at sixty-six. Not bad.

Beware, however: the opposite problem also exists, that of geniuses living longer than their genius, a problem that, frankly, we do not know how to solve. This is the case of the American writer Jerome D. Salinger, whose artistic productivity lasted about a decade, from the age of thirty-two—when he published *The Catcher in the Rye*—to forty-two—when he published *Franny and Zooey*. After that, he retired to private life—indeed, he became a recluse—living to the age of ninety-one. Also reaching the age of ninety-one was the Finnish composer Jean Sibelius, who, however, stopped composing thirty years before his death, refusing to answer those who questioned him about his artistic silence. And what about the French poet Arthur Rimbaud, who stopped writing at the age of twenty-one, only to devote the rest of his life—he died at thirty-seven—to travelling and trading. To conclude this sub-theme, we certainly do not wish to die young; on the contrary: do your best to become long-lived and highly productive teachers.

5.8   Isolation or Intense Social Life?

And here, we come to another well-known stereotype, that of the solitary genius. This is a cliché that competes in popularity with that of the mad genius, and in this, as in the latter, there is some truth. We have to make some distinctions, however, particularly with regard to the way in which this solitude is understood: on the one hand, we can, in fact, ask ourselves whether producing works of genius requires especially intense solitary work, which would therefore represent a necessary constraint for the genius; on the other hand, we have to consider whether, from a psychological point of view, being a genius is connected in any precise way to a naturally solitary personality. Is there a connection, that is, between genius, introversion, obsession with work, and loneliness? And, so we can begin to contemplate this, Simonton immediately provides us with a list of genius personalities—one that you can also easily obtain with Google or Wikipedia—who would have been historically lonely: not only Newton, but also Frederic Chopin, Charles Darwin,

Bob Dylan, Albert Einstein, Bill Gates, Jimi Hendrix, Piet Mondrian, Ayn Rand, Jerome D. Salinger, Aleksandr Solzhenitsyn, and Nikola Tesla. Making lists of this kind mostly gives us stories to tell; what we need, however, is a rigorous analysis of whether, normally, geniuses are more reclusive and introverted than the average general population. So, let us turn again to one of the pioneers of the study of genius, Raymond B. Cattell, and take a look at his work on the subject. In addition to his groundbreaking work, Cattell also had a habit of coining neologisms that were rather difficult to parse. Here are a couple of them: "schizothymic" and "desurgent." In substance, according to his research, top scientists tended to be isolated, inwardly focused, critical and precise, as well as sceptical—hence, schizothymic; beyond that, they were distinguished by introspection and solemnity of manner—desurgent. Cattell's analyses were mainly based on posthumous biographical assessments. According to the above-mentioned research by Ravenna Helson, the female mathematicians she studied tended to prefer solitary pastimes. The same conclusions were drawn by Anne Roe with regard to the sixty-four eminent scientists she studied. According to the American researcher, this tendency is mainly rooted in a childhood that was also solitary and not very social.[29] The fact, however, that creative geniuses are introverted does not mean that they have a resigned and submissive relational style; on the contrary, they would often tend to be very autonomous, and to take the lead if there are precise objectives—linked to their work—to be achieved. This is the case of Einstein, for example, who normally preferred to work alone, but could also find himself working with others, in the event, for example, that he was struggling with mathematical questions beyond his own competence. In short, social relationships would only be cultivated in an instrumental sense, as a means to continue the work of creative genius. Anne Roe also notes that a typical characteristic of the subjects she observed is a very strong absorption in their own work, which takes priority. And why all of this work? Answering this question leads us to delve not only into the psyche of geniuses, but into the human psyche in general; what Roe says, in fact, applies not only to scientists, but to artists of all kinds, and indeed to any human being who has cultivated at least a modicum of awareness of their own mortality. "Each works hard and devotedly at his laboratory, often seven days a week. He says his work is his life, and has few recreations. ... [These scientists] have worked long hours for many years, frequently with no vacations to speak of, because they would rather be doing their work than anything else." In short, they do what they love deep down, knowing that their time is limited, in the conscious

---

[29] Roe, Anne (1952). *A psychologist examines 64 eminent scientists,* Scientific American, 187(5), p. 22.

or unconscious hope of producing something great, which will perhaps make them go down in history, in a form of surrogate immortality. Obviously, if we dive into the subject of motivation for action, we will end up getting off track, as well as getting lost in the vast thicket of different schools of psychology and all the theories relating to human motivation. Better, then, to try to work on the second part of our equation, the one concerning the peculiar sociality of geniuses. For, if we analyse any work of genius, be it a novel, a statue, or a scientific theory, the first thing we will do is to try to situate it within a precise cultural-historical context; this is simply another way of saying that, far from being truly completely isolated, creative geniuses have in some way communicated constantly with their peers, or rather with those among their peers who are akin to them in terms of interests, passions, and whatnot. We need only think of Mihaly Csikszentmihalyi's work: as you may recall, the psychologist maintains that true creation stems from the intertwining of individual, domain and field, and would therefore not be something purely psychological. Not only that: such a theoretical approach would show us how the work of creative geniuses, no matter how isolated, could never be divorced from relations with the surrounding world, and, in particular, with kindred spirits. This, then, is the answer to our initial question: the genius takes full advantage of the network of contacts available within a certain social context, and it could not be otherwise. Let us consider, for example, the case of Newton, one of the most reclusive geniuses: a friend of Edmund Halley, their relationship was based mainly on the elaboration of the *Principia*; let us also not forget that the scholar took part in several disputes, including one with Leibniz on the origins of infinitesimal calculus. In short, although Newton was psychologically a recluse, this did not prevent him from overcoming this attitude when the influence of his scientific work was at stake.

5.9   Genius, Race and Gender

Here, we come to the thorniest topic of all, that of the relationship between genius, gender and race. Don't worry, though, we will be politically correct. Not least because there is not much to say on the topic in question that has not already been said before, and we will limit ourselves here to summarising the current state of the debate on the subject. It all starts—as always—with Galton, and the fact that, as a child of his time, the scientist argued for the mental inferiority of women and non-white races. We contemporaries move in a very different cultural context, in which racism and sexism are condemned and banned from official culture, as are all philosophical-scientific doctrines that support such attitudes. The writer, too, obviously takes this view of things, although we will avoid entering into technical debates on the existence or non-existence of human races—in anthropological circles, in fact, it is denied that race is a meaningful concept—and on the existence or non-existence of biological differences between men and women—a subject that is still hotly debated. Let us therefore take up Galton again, as well as Franz Joseph Gall and his cranioscopy. It was he, in fact, who 'opened the dance' on the subject of measuring skulls and, especially, comparing skulls belonging to individuals of different races. In short, while it is today used as a term to indicate an attitude of discrimination towards people of different ethnicities, the word 'racism,' in the nineteenth century and the first half of the twentieth century, indicated a body of pseudo-scientific doctrines that believed not only that human races existed, but that they were also arranged hierarchically, with white Europeans at the top of the pyramid. And such a view of things could not help but infiltrate the discourse on genius conducted at the time. The same applies to the question of the relationship between genius and gender. Let us therefore start with the question of racism. Galton devotes an entire chapter to the ranking of intellectual abilities on a racial basis, making the specific argument that the peoples of Sub-Saharan Africa would be inferior to whites; this does not place the latter, and, in particular, the Anglo-Saxons, at the top of the human intellectual pyramid. Such a role would, in fact, belong to the inhabitants of Ancient Greece, but since this group is situated in the past, the role falls—by default—to the Anglo-Saxons.

In the years following the Second World War, eugenics was abandoned fairly quickly, and today, as we have said, the world of official culture has undergone a radical paradigm shift with regard to the concept of race. It is considered either meaningless—a mere social-cultural construct that does not correspond to any biological reality—or as a real but biologically insignificant concept. And, in any case, it is a concept with blurred, vague boundaries, so that it is not possible to identify truly homogeneous racial groups upon which to then conduct investigations into intelligence and genius. And even if we were able to do so, we would end up finding that, in any case, IQ varies within one racial group as much as it varies within other racial groups. Moreover, it is impossible to prove that a certain intellectual difference between any two different racial groups really depends on a different biological substratum. For instance, if we take a look at the average IQ of the different American sub-groups—whites, blacks, Asians, and so on—we will see that the relative differences that emerged in the past have gradually smoothed out, due to, for instance, generally better nutrition or more widespread educational opportunities. Note, then, one thing, namely, that we have avoided equating the concepts of race—an idea of an essentially biological nature—and that of ethnicity—which is, in fact, a cultural concept. In fact, it refers to groups of people united by a common history, a common language and common traditions, in short, all exquisitely cultural notions, not biological ones. It is clear that if race does not play a role in IQ and genius, ethnicity plays even less of a role. In short, it is now commonly accepted that differences in genius productivity depend on the presence or absence of access to economic and educational opportunities.

The fact remains, however, that, in human civilisation, the bulk of cultural contributions come from male individuals, and this introduces the second issue, that of genius and gender. Now, at present, the issue is rather thorny, especially as far as the academic world is concerned, divided between those who maintain that gender is something different from sex and constitutes a social construct and those who instead defend a conception closer to biology. We, for our part, simply choose to point out the causes commonly associated with this visible disparity, starting with different educational methods—i.e., a certain educational style and related expectations for males, a completely different style and expectations for females. This difference is present in almost every major civilisation on the planet. If we then want to get more specific, let us recall the active discrimination against women who have tried to pursue creative goals throughout our history. Marie Curie was not admitted to the French Academy despite two Nobel Prizes and did not obtain a stable academic position until the death of her husband. Anne Sexton wrote her poetry in her spare time from household chores and raising two children. And let us not forget Amantine Lucile Aurore Dupin de Francueil, a French writer and journalist who signed herself as George Sand, and Mary Ann Evans, an English writer and poet who signed herself as George Eliot: both authors chose to use male pseudonyms to avoid prejudice. In conclusion, the spirit of the times now tells us that race and sexual gender are no impediment to the development of genius creativity, and that inequalities depend on discriminatory social structures, lack of economic opportunity and training. This being the case, in the future, we should see—as society changes—a levelling out of ethnic and gender differences in creative genius. Everything we have said above should, however, be distinguished from the nature-or-nurture issue, in the sense that to hold that genius is biological in origin does not necessarily mean that there is a racial or sexual difference in its distribution, but only that it is rooted in biology rather than culture.

5.10   Education and Marginality

What is the relationship between genius and the education system? Here, again, we are in the presence of a commonplace, namely, that formal education would be an impediment to genius, or, at any rate, would not help it at all. Proverbial, in this regard, are the words of Albert Einstein, who, on one occasion, said that "it is, in fact, nothing short of a miracle that the modern methods of instruction have not yet entirely strangled the holy curiosity of inquiry; for this delicate little plant, aside from stimulation, stands mostly in the need of freedom; without this it goes to wreck and ruin without fail. It is a very grave mistake to think that the enjoyment of seeing and searching can be promoted by means of coercion and a sense of duty." In particular, the great scientist had it in for tests: "One had to cram all this stuff into one's mind for the examinations, whether one liked it or not. This coercion had such a deterring effect on me that, after I passed the final examination, I found the consideration of any scientific problems distasteful to me for an entire year." With such an attitude, it is easy to imagine the surprise of his professors when Einstein reached the heights of contemporary physics. Echoing Einstein is none other than Darwin himself, who recounts how the geology and zoology lectures he found himself attending in Edinburgh were enormously boring: "The sole effect they produced on me was the determination never as long as I lived to read a book on Geology, or in any way to study the science." Luckily for us, the scientist later changed his mind. Be careful, however: these are only anecdotes, and we know for a fact that many eminent people were more than diligent in their studies. Marie Curie, for example, who was two years ahead of her peers in primary school, or Freud, who was top of his class in grammar school, or even Robert Oppenheimer, who graduated with honours from Harvard. Anecdotes are not enough, however: you need data. And here they are: a survey of the university grades of the Fellows of the Royal Society revealed a very weak relationship between the scholastic performance of the scholars in question and their subsequent professional success.[30] This lack of correlation between scholastic performance and subsequent creativity is particularly evident in the field of art. Not only that, but creative genius is not even linked to the attainment of a degree, as we can see from the cases of Newton and Darwin, who barely made it to a bachelor's degree, as well as some—like Michael Faraday—who did not even go to university. To

---

[30] Hudson, L., *Undergraduate academic record of Fellows of the Royal Society*, Nature, 182,1326, (1958).

this must be added, however, that those who achieved scientific or artistic eminence without going to school still had to undertake a tough course of self-education.

For example, there's Darwin, who engaged in a profound programme of self-education through extensive reading and frequent conversations with eminent scholars. "I consider that all that I have learned of any value has been self-taught," he said. And then there is the famous phrase attributed to Mark Twain: "I have never let my schooling interfere with my education." However, let us not exaggerate: after all, a minimum of basic skills and knowledge, acquired through a formal education programme, are undoubtedly indispensable for anyone, including geniuses, hence the latter's ambiguous relationship with the school system. However, if this system becomes rigid, it could have deleterious effects on individual creativity. And this is why many of the aforementioned creatives have supplemented formal education with a self-education programme. Not only that, but—Simonton ventures—there is the possibility that the most revolutionary scientists tend not to be particularly brilliant in academia. Let's cite the ubiquitous Einstein, who obtained his doctorate while working full-time at the Swiss patent office, therefore without the possibility of receiving a university education. And he got it by submitting the least important of his four papers, while the most important ones, which allowed him to shake up contemporary physics, he kept to himself.

And now, let us introduce the last topic of the chapter, the—very interesting—topic of marginality. Let us start with the Termites: they were, in fact, representative of the majority in the most classical sense of the term, i.e., they were mostly WASPs—White, Anglo-Saxon and Protestant. This fact clashes with the view that human creativity could only be fostered in those who were in some sense marginal persons, i.e., belonging to a minority that was somewhat isolated in a social sense. The sociologist Robert Park, for instance, emphasises the role that immigrants can play in creative cultural change,[31] i.e., situations in which one single person is torn between two different cultural groups. For his part, the famous historian Arnold Toynbee coined the term 'creative minority,' by which he indicates those groups of people who withdraw, i.e., separate, from a certain majority culture, only to return to it, bringing with them cultural innovation.[32] Marginality thus has a privileged link with belonging to a certain ethnic group. Donald Campbell points out in this regard that "persons who have been uprooted from traditional culture, or who have been thoroughly exposed to two or more

---

[31] Park, Robert E., *Human migration and the marginal man*, American Journal of Sociology, 33,881–893, 1928.
[32] See Toynbee, Arnold J., *A study of history*, Oxford University Press, Oxford 1946.

cultures, seem to have an advantage in the range of hypotheses they are apt to consider, and through this means, in the frequency of creative innovation."[33] According to this hypothesis, the ethnically marginalised person should manifest a greater richness of associations of ideas and a more divergent way of thinking. Some data seem to confirm the correlation between marginality and increased creativity: for example, research has shown that, in the United States, 19 per cent of the eminent persons examined in this study were first- or second-generation immigrants.[34] The most exemplary case of the relationship between ethnic marginality and scientific and artistic creativity is the Jewish people. Although very small in numbers, this people has provided the modern world with an enormous number of intellectuals, artists, scientists, and so on.[35] A particularly enriching family environment, bilingualism or multilingualism, a propensity for mobility, and thus openness to different mentalities and influences, could explain this success.

To this, we add another type of marginality, the disciplinary one: in fact, it is not to be excluded that, at least in certain cases, a decisive contribution in a certain discipline may be made by people who are culturally external to it. This is the case, for example, with the end of the dinosaurs, whose extinction by the famous asteroid—which tormented generations of palaeontologists— was explained not by a specialist in the field, but by a nuclear physicist, Luis Alvarez.

Let us conclude the chapter with a final general reflection: whether genius is the result of nature or nurture, it seems obvious that a shrewd society should do its best to foster the birth and development of individual genius, first and foremost, by avoiding hindering the latter with social and economic stakes, in the hope that economic prosperity and automation of the most monotonous and non-creative jobs will allow more and more people to discover and cultivate their genius.

---

[33] Campbell, Donald T., *Blind variation and selective retention in creative thought as in other knowledge processes*, Psychological Review, 67,380–400, 1960.

[34] Goertzel, M. G., Goertzel, V., and Goertzel, T. G., *Three hundred eminent personalities: A psychosocial analysis of the famous*. Jossey-Bass, San Francisco 1978.

[35] Arieti, Silvano, *Creativity: The magic synthesis*, Basic Books, New York 1976, pp. 325–326.

6

Forcing the Skies

6.1 Towards a Psychology of Science

If the great geniuses of art manage to amaze us with works that give us a glimpse of a surreal beauty that, as Dante would say, "transhumanises us," the great geniuses of science offer us a product that is apparently different, although, in reality, connected to the former: in particular, they enable us to pierce the veil of appearances, in short, to realise, at least to the slightest degree, Stephen Hawking's dream, that is, "to know the mind of God." It is to the latter that we shall now turn, trying to understand what distinguishes a scientist of genius from a mediocre one, and what the path to scientific excellence may be. Here, too, we shall draw on the work of numerous researchers and, in particular, that of Simonton, an author who has tried, among other things, to develop a true psychology of science, to be placed alongside the philosophy, history and sociology of science.

We already know the American psychologist's theory: the mind is a random generator of ideas, which are then subjected to a Darwinian process of selection and retention of what works. Then, in the case of scientific research, the talented scientist finds him or herself caught between two antagonistic forces, as, on the one hand, they must learn and, above all, preserve the fruits of past research and, on the other hand, they must innovate. This contrast

© The Author(s), under exclusive license to Springer Nature
Switzerland AG 2023
R. Manzocco, *Genius*, Springer Praxis Books,
https://doi.org/10.1007/978-3-031-27092-5_6

between traditionalism and iconoclasm has been christened by the philosopher of science Thomas Kuhn "the essential tension."[1] Even in the case of
scientific research—indeed, especially in this case—a process of random variation and selection acts in the mind of a scientist, Simonton assures us. In
this case, it is not organisms or genes that undergo selection, but what the
scholar calls "mental elements." Such elements—ideas, images, concepts—do
not even have to be conscious. On this subject, Einstein said: "It seems to
me that what you call full consciousness is a limit case which can never be
fully realised. This seems to me connected with the fact called narrowness of
consciousness."[2] It is not easy, however, Simonton warns us, to define the term
'randomness,' which, in the common—i.e., reductionist and mechanistic—
conception of part of contemporary science, simply indicates our degree
of ignorance—in essence, one would define something as 'random' when
one does not know its causes. The scholar cuts the bull's eye off the head,
contenting himself with admitting the existence of a vast number of possible
permutations, all characterised by a low but non-zero degree of probability.
To put it another way, we assume that the combinations of mental elements
are many and all more or less equally probable. In short, Simonton envis-
ages for us a real 'mental chemistry,' in which the various elements combine
randomly into unstable permutations—confused, chaotic ideas—which he
calls "aggregates," and stable permutations—Cartesianally clear and distinct
ideas -, renamed "configurations." Of course, there is no clear distinction
between aggregates and configurations; they simply represent the two ends
of a continuum. Our thoughts would sometimes be very clear, sometimes
confused, along with everything in between.

In the minds of scientists—or, for that matter, of all people—a process of
progressive self-organisation of these mental elements would also be taking
place, whereby—at least in certain cases—we human beings would find
ourselves experiencing cognitive events that reduce the level of entropy—i.e.,
chaos—in our minds. We are in the presence of what Maslow called "peak
experiences," or what, more simply, we might call "revelations." Not necessarily in the religious sense: revelations therefore in the sense of moments
of extreme clarity, when a certain situation or existential condition suddenly
appears to us in a new, clearer light. Regarding science more specifically,
it constitutes, for Simonton, a system of configurations that progressively
self-organises. This should not, however, be confused with the dispassionate

[1] Kuhn, Thomas S., *The Essential Tension: Selected Studies in Scientific Tradition and Change*, University of Chicago Press, Chicago 1977.
[2] Quoted in: Hadamard, Jacques, *The Psychology of Invention in the Mathematical Field*, Princeton University Press, Princeton 1945, p. 143.

search for truth: in fact, the confirmation bias—i.e., the natural tendency to pay attention only to what confirms our assumptions—operates in all human beings, scientists included,[3] and indeed, the scientists most respected by their colleagues would be precisely those who strive to prove their most cherished hypotheses.[4] From a psychological point of view, the Popperian idea that scientists would formulate hypotheses and then patiently try to dismantle them in every way is just a myth.

So: the human mind would be a machine that generates random permutations of ideas and gradually builds up an increasingly coherent vision of the world and life and, in the case of the scientist, a specific vision of the discipline that interests them. In us, therefore, there would be a vision of things that is self-organising, as we live experiences and arrange them in this personal internal scheme, always under construction and always influenced by our cognitive biases. But that is not enough: in the case of science, mental configurations have to be communicated, and for this to happen, they have to be translated into formulations that can be understood by most. So, a certain scientific intuition, no matter how ineffable, must be translated into what Simonton calls a communication configuration.

The next stage then consists, for the scientist, in getting his or her communication configuration accepted, through personal influence and persuasion, i.e., speaking the language and meeting the acceptance criteria of the people they want to convince. If things go well, we then move on to the stage of socio-cultural preservation, i.e., the vision proposed by the individual scientific genius is incorporated into the generic vision accepted by the scientific community.

We have already shared a few anecdotes confirming the validity of the Simontonian hypothesis in the chapter on the Darwinian nature of genius; now, let us add a few more. First of all, as usual, we turn to Hadamard,[5] who points out how many mathematicians report that, in the early stages of a discovery, there is a predominance of visual images and, at times, kinesthetic sensations in their minds. And, because this process of discovery does not see much participation by verbal thought, the ideas in question are not articulated, but often chaotic. And, when they do participate, words do so in atypical ways. This is the case, for example, with Galton, who, at one point, observes that, while he is thinking, meaningless words may appear

---

[3] Mahoney, Michael J., *Scientist as Subject: The Psychological Imperative*, Ballinger, Cambridge MA 1976.
[4] Mitroff, Ian, *Norms and Counter-Norms in a Select Group of the Apollo Moon Scientists: A Case Study of the Ambivalence of Scientists*, American Sociological Review, 39, 579–595.
[5] See Hadamard, Jacques, op. cit., 1945.

"as the notes of a song might accompany thought."[6] Of course, it also depends on the discipline we are considering, and some disciplines—such as the social sciences—are more dependent on linguistic processing than others. If, however, we remain in the field of natural sciences, then the objects that specialists in these fields deal with are natural phenomena of a non-verbal nature. And it is therefore logical that, in order to obtain new permutations, the scientific genius must go beyond words. And here, we can hazard a guess, namely, that the more original a permutation is, i.e., the further from linguistically established common sense, the more difficult it will be to put into words.

If we want, we can also add a bit of etymological analysis to the mix of our reasoning. To this end, we can mention that the German/British philologist Max Muller pointed out that the Latin verb *cogito*—meaning 'to think'—originally meant 'to shake together,' a definition that fits in well with Simonton's conception; while St. Augustine apparently maintained that 'intelligo,' the Latin root of the word 'intelligence,' meant 'select among,' i.e., the second phase of the Simontonian process, selection.[7]

6.2   Incubation

Remaining in the anecdotal sphere, let us cite the work of the well-known British political activist and social psychologist Graham Wallas, who—in his work *The Art of Thought*[8]—illustrates one of the earliest models of the creative process—which later became very popular in the academic sphere—structured into four phases: preparation, incubation, illumination and verification. And Wallas bases his theoretical model on the anecdotal accounts narrated by Helmholtz in his autobiography.[9] Even for Poincaré, sudden intuitions would be "a manifest sign of long, unconscious prior work."[10] In practice, the incubation phase—operated by what he calls the "subliminal self"—would complete the work begun with the preparation, and would represent—according to Simonton—the place where the process of chance varation would take place. And Hadamard seems to agree: "…it has been necessary to construct the very numerous possible combinations,

---

6 Quoted in: Ibid, p. 69.
7 Both cited in: Ibid, p. 29.
8 Wallas, Graham, *The Art of Thought*, Solis Press, Poole (Dorset) 1926.
9 Helmholtz, Hermann von, *An autobiographical sketch*. In: *Popular Lectures on Scientific Subjects, Second Series*, Longmans, New York 1898, pp. 266–291.
10 Poincaré, Henri, *The Foundations of Science*, Science Press, New York 1921, p. 389.

among which the useful ones are to be found. It cannot be avoided that this first operation take place, to a certain extent, at random, so that the role of chance is hardly doubtful in this first step of the mental process. But we see that that intervention of chance occurs inside the unconscious: for most of these combinations-more exactly, all those which are useless-remain unknown to us."[11] And these episodes—incubation followed by revelation—are said to be quite common: according to an old study from 1931, conducted by two American researchers—Washington Platt and Ross Baker-, only 17 per cent of the 232 eminent scientists interviewed never went on to have a scientific revelation.[12] To this, Simonton makes an interesting addition: "The more original the illumination is, the longer the incubation, on the average, will be. That is, if the elements that make up the chance configuration are not related to one another, so that they can be connected only by rather long and strange sequences of associations, then the odds are greater that the permutation mechanism must run longer before the stable combination will emerge."[13] There are certainly points in common between the theories on scientific creativity set out so far and one further approach, that of the so-called "bisociation" advocated by Arthur Koestler. In particular, the Hungarian writer and journalist devoted a book that has become very famous in relation to the subject, *The Act of Creation*.[14] In it, Koestler reviews and compares imaginative and inventive processes in the most diverse fields—in the arts, sciences and even humour—with the aim of developing a general theory of human creativity. According to him, all these phenomena share a common pattern that he christened "bisociation," which consists of a mixture of elements drawn from unconnected "matrices of thought." The matrices in question would denote "any ability, habit, or skill, any pattern of ordered behaviour governed by a 'code' of fixed rules."[15] Additionally, according to Koestler, bisociative creative intuitions would occur after a long and intense conscious effort at problem-solving, during a period of relaxation, at a time when rational thought is put in brackets. However, one of Simonton's criticisms of Koestler's theories is that they lack testable statements, i.e., they are essentially incomplete.

While we're on the subject of sudden revelations and insights, let's take a look at those that occur in a dream, while sleeping or under the influence of

---

[11] Hadamard, Op. Cit., p. 28.

[12] Platt, Washington and Baker, Ross, A., *The relation of the scientific 'hunch' to research*, Journal of Chemical Education, 1931, vol. 8, p. 1969.

[13] Simonton, Dean K., *Scientific Genius. A psychology of science*, Cambridge University Press, Cambridge 1988, p. 33.

[14] Koestler, Arthur, *The Act of Creation*, Hutchinson, London 1964.

[15] Ibid. p. 38.

drugs. There is one thing strikes us about them, and that is that they are very few indeed. They can be counted on the fingers of one hand. In particular, let us recall Kekulé's aforementioned dream about benzene atoms arranged in a circle; regarding it, the scientist declares that: "I fell into a reverie, and lo! the atoms were gambolling before my eyes."[16] And then, we have the case, unrelated to scientific research, of Coleridge and his opium-produced dream. Little, indeed, to conclude that drugs are of any help to creativity. It is more likely that normal states of consciousness are better suited to scientific discovery than those altered by psychotropic substances. Not to mention 'false positives,' ideas that, while they seem brilliant or ingenious during an altered state of consciousness, later—in a lucid mind—no longer seem so. In this regard, William James tells of his experiences with nitric oxide, during which he would find himself meditating on apparently profound concepts, which turned out to be rather silly once he regained normal awareness.[17]

6.3   For a Psychology of Scientists

Let us now take a look at the character traits of the geniuses of science, not least because, while eminence is related to IQ to a certain extent, personality probably has a lot to do with it. We won't even attempt to go into the contemporary debate on personality theory, that is, the area of psychology that seeks precisely to define and classify psychological traits. All you need to know is that, at present, there is opposition between those who believe that psychological traits—or, at least, some of them—are relatively stable and those who, on the other hand, believe that everything depends on the context, so that personality per se, as a concept, would be meaningless. We adhere to the first school of thought, and believe that at least some traits are sufficiently stable and linked to recurring behaviour. And so, we will try to identify the psychological differences—assuming there are any, of course—between scientific geniuses and people who are simply talented but not brilliant (or even mediocre). This is a distinction that Simonton wants to connect to his theory of creativity, based on the following reasoning: a person who produces more ideas—or, to put it in his language, a person who generates more random permutations—is likely to have other character traits related to this characteristic. We therefore have two aspects to consider when assessing the psychology of science geniuses: cognitive style and motivation.

---

[16] Quoted in: Findlay, Alexander, *A Hundred Years of Chemistry*, Duckworth, London 1948, p. 37.
[17] See James, William, *The Varieties of Religious Experience*, Longmans, London 1902.

Let us start with the first, and see where the American scholar's line of thought takes us. And here, having previously been thrown out the door, the IQ comes back in through the window: to be a top scientist, the IQ does, in fact, count for something: in the case of PhD students in physics, for example, it would be around 140 points. Which is not bad at all, if we calculate that the typical IQ of the undergraduate student in the first years of university is around 118.[18] Let us also bear in mind Cox's analysis, according to which the IQ of the historical figures she studied—with all of the limitations already mentioned—ranged from Darwin's 156 to Newton's 190. We are now familiar with the theory promoted by Simonton and many others: the IQ represents a threshold that must be exceeded in order to be creative; once exceeded, however, this value loses its importance, so that one can be extremely intelligent and not very creative, or moderately intelligent but very creative. Still, in relation to the intellect of the brilliant scientist, it must be structured in a certain way. In particular, it must be characterised by verbal fluency, originality, breadth of interests, flexibility and independence of judgment.[19] It also seems that the presence of a strong interest in an area of study outside that within which one normally works is correlated with scientific success.[20,21] A common trait of great scientists seems to be what British physicist and historian of science Derek J. de Solla Price called "mavericity," which he defined as the ability to connect ideas in unexpected ways.[22] At this point, Simonton introduces a distinction that is very important to us, namely, that between "intuitive genius" and "analytical genius". These are two labels he uses to denote the cognitive styles typical of eminent scientists. Thus, "the 'intuitive genius' has many elements linked by numerous infraconscious but behaviourally and emotionally active associations, the 'analytical genius' has a comparable number of elements linked by a smaller set of conscious and ultraconscious cognitions and habits—therein yielding the distinction between intuitive and analytical styles of thought."[23] These are two different cognitive styles: the analytical genius has a well-structured vision of reality,

[18] See, for example, Roe, Ann, *The Making of a Scientist*, Dodd, Mead, New York 1952.

[19] Barron, Frank, *The needs for order and disorder as motives in creative activity*, in: *Scientific Creativity*, ed. C. W. Taylor & F. Barron, pp. 153–160, Wiley, New York 1963.

[20] Simon, Rita J., *The Work Habits of Eminent Scholars*. Sociology of Work and Occupations, 1(3), 327–335, 1974.

[21] Andrews, Frank M.; Aichholzer Georg, *Scientific Productivity, the Effectiveness of Research Groups in Six Countries*, Cambridge University Press, Cambridge 1979.

[22] Price, Derek, *Little Science, Big Science*, Columbia University Press, New York 1963, p. 107.

[23] Simonton, Dean K., *Scientific Genius. A psychology of science*, Cambridge University Press, Cambridge 1988, p. 46.

articulated but, precisely for this reason, rigid, while the intuitive genius—which, in some respects, represents a phase in the life of the genius scientist that precedes the analytical one—has a more disordered form of thinking, a vision of things, but, precisely for this reason, more open to connections between even quite remote elements. As the scientific genius proceeds in their intellectual career, their vision of things will become increasingly self-organised, acquiring completeness and organicity, but, at the same time, reducing its associative capacity. If, in fact, the elements in the scientist's mind always remain the same, it is clear that, as time passes, they will become more and more organised, and he or she will go from being an intuitive genius to being an analytical genius.

Let us now take a look at the motivation of genius scientists. In particular, let us make a distinction between the psychological traits of reinforcement—the drive to do research—and the psychological traits of inhibition—the drive to give up other things in order to devote oneself exclusively to science. As far as reinforcing motivation is concerned, let us remember that the successful scientist is full of energy and is a hard worker,[24] is fascinated beyond measure by research[25] and devotes an inordinate amount of time to it,[26] is ambitious and persistent[27] and is completely absorbed by work.[28] As for inhibitory traits, we often have renunciation of interpersonal interactions, politics and social activities in general[29,30]—better to be alone, pondering the ultimate nature of reality—and little dedication to teaching.[31] But we do not stop there: in a study dedicated to the psychological profile of some eminent scientists, Raymond B. Cattell highlighted how many of them are characterised by schizothymic and desurgent traits.[32] And, as an example, Cattell cites the case of Henry Cavendish—an important physicist and chemist who lived in the

[24] Bloom, Ben S., 1963. *Report on creativity research by the examiner's office of the University of Chicago.* In: *Scientific Creativity: Its recognition and development*, ed. C. W. Taylor & F. Barron, pp. 251–264, Wiley, New York 1963.

[25] Blackburn, Robert T.; Behymer Charles E. and Hall, David E., *Correlates of Faculty Publications,* Sociology of Education Vol. 51, No. 2 (Apr., 1978), pp. 132–141.

[26] For example: Gaston, Jerry, *Originality and competition in science: A study of the British high energy physics community,* University of Chicago Press, Chicago 1973, p. 51.

[27] Rushton, J. Philippe, Murray, Harry, & Paunonen, Sampo, *Personality, Research Creativity and Reaching Effectiveness in University Professors.* Scientometrics, 5, 93–116, 1983.

[28] Roe, Ann, Op. Cit., p. 25.

[29] Roe, Ann, Ibid.

[30] Terman, Lewis M., *Are Scientists Different?*, in: Scientific American, vol. 192, pp. 25–29, 1955.

[31] Fulton, Oliver, & Trow, Martin, *Research activity in American higher education,* Sociology of education, 47, pp. 29–73, 1974.

[32] Cattell, Raymond B., *The personality and motivation of the researcher from measurements of contemporaries and from biography.* In: *Scientific Creativity: Its recognition and development*, ed. C. W. Taylor & F. Barron, pp. 119–131, Wiley, New York 1963.

eighteenth and nineteenth centuries and who, among other things, discovered hydrogen. Cavendish, "when dragged to a state function and about to meet some distinguished … foreign scientists, broke away and ran down the corridor, squeaking like a bat."[33] Admittedly, Cavendish represents a bit of an extreme case: he addressed only a few words to men and did not speak at all to women, he communicated with his maids only by written notes, and he equipped his own house with a special entrance separate from the main one in order to enter and leave without having to meet anyone. All of this is to say that scientific geniuses would tend to be much more introverted people than the average person, albeit usually in less extreme ways.

Even more important than introversion is intrinsic motivation: in other words, great scientists, however tempted they may be by fame, money and power—i.e., what we might consider extrinsic, socially determined motivations external to pure scientific research—will always be motivated far more by intrinsic interest in the discipline they study, in the mysteries it still conceals and that await to be unveiled.[34] If one wastes time fantasising about social success, one will steal important energy and time from the reflection that then leads to exceptional discovery. In short, there can be no homologation to dominant values, be they social or economic success, if you want to be a genius scientist: rather, you need full existential autonomy from social pressures and absolute dedication to the object of your research.

And what about the so-called 'Big Science,' i.e., all of those fields of research that are often well-funded and based on the use of complex machinery, and even very large teams of researchers? Such as high-energy physics and biomedical research? These are fields in which a single piece of research sometimes sees the collaboration of dozens of people, in which the proper management of interpersonal relations, the need to be politically astute and the simple pressure to publish play a key role, and leave no room for the behavioural eccentricities of some of the great scientists of the past. Sometimes, a particular research group reaches such a level of success and prestige that it can afford a certain degree of relaxation, thus making the interference of external factors irrelevant. And this, in turn, allows for free and relaxed group discussions and productive exchanges of opinions—what is nowadays called 'brainstorming,' although, traditionally, this term indicates a specific technique in which the members of a certain group expound their ideas spontaneously and uncensored. Despite all of this, empirical research has shown that individual brainstorming appears to be much more efficient

---

[33] Ibid. p. 121.
[34] Amabile, Teresa M., *The Social Psychology of Creativity*, Springer, New York 1983.

than collective brainstorming, both in terms of the quantity of ideas generated and their quality.[35,36,37] Better to do it yourself then, and when this is
not possible, keep in mind that the best ideas often come when you brain-
storm on your own. Indeed, it is very likely that the brilliant ideas, the ones
that change the cards on the table in a certain field of research, are almost
exclusively born during periods of solitary meditation.[38] And Simonton rein-
forces his thesis by also invoking Einstein and Newton. The former is said
to have said: "I am a horse for a single harness, not cut out for tandem or
team-work; for well I know that in order to attain any definite goal, it is
imperative that *one* person should do the thinking and the commanding."[39]
As for Newton, the poet William Wordsworth called him "a mind for ever/
Voyaging through strange seas of thought, alone."[40]Simonton, at this point,
puts forward a hypothesis, albeit without pushing it too far: what if, in trying
to identify the personality of the scientific genius, we had actually highlighted
the characteristics of genius *tout-court*? Could we say that there is a 'generic
genius'? A genius could pursue beauty or truth, but, in any case, they would
need exceptional abilities and a psychological profile that would lead them to
dedicate themselves to their work for many years with superhuman dedication. Better not to exaggerate, however, says the American scholar, with this
'generic genius': the common point between art geniuses and science geniuses
seems to be precisely this extreme dedication, seasoned with stainless patience.
However, it is better not to exaggerate the differences either: according to
Simonton, there is a fair amount of overlap between the cognitive styles of
artists and scientists. Thus, the two categories would be close enough to
be lumped into the broader category of 'generic genius,' without, however,
forgetting the differences in method and objective. And, in this regard, the
scholar tells us that: "Wilhelm Ostwald, the 1909 recipient of the Nobel Prize
for Chemistry, distinguished between two varieties of scientific researchers,
namely, the classicists, who are systematic perfectionists, and the romanticists,

---

[35] Bouchard, Thomas J. Jr. and Hare Margaret, *Size, performance, and potential in brainstorming
groups*, The Journal of applied psychology, vol. 54:1, pp. 51–55, 1970.

[36] Dillon, Peter C.; Graham William K.; Aidells, Andrea L., *Brainstorming on a "hot" problem: Effects
of training and practice on individual and group performance*, Journal of Applied Psychology, vol. 56,
pp. 487–490, 1972.

[37] Dunnette, Marvin D.; Campbell, John; Jaastad, Kay, *The effect of group participation on brainstorming effectiveness for 2 industrial samples*, Journal of Applied Psychology, 47, pp. 30–37,
1963.

[38] Cf. Pelz, Donald Campbell, and Andrews, Frank M., *Scientists in Organisations: Productive Climates
for Research and Development*, Institute for Social Research, University of Michigan, Ann Arbor 1976.

[39] Quoted in: Sorokin, Pitirim, *A Long Journey; the Autobiography of Pitirim A. Sorokin*, CT: College
and University Press, New Haven 1963, p. 274.

[40] Wordsworth, William, *The Prelude*, Penguin, New York 2010, p. 122.

who have such a profundity of ideas that they fail to develop them fully. By the same token, artistic creators might be similarly differentiated, the classicists favouring precision, order, and stability, and the romantics, expression, richness, and novelty."[41]

6.4 Massive Industrial Production of Ideas

It is now the turn of another characteristic that distinguishes scientific geniuses from their mediocre counterparts: the immense productivity of the former. There is, in fact, a rather erroneous cliché that great minds in science contribute a great discovery or theory, for which they will mainly be remembered. But let us look at the various factors in order. And let us start with the age of the first scientific contribution, which is generally relatively low. In fact, Simonton tells us that "creative output [tends?] to increase relatively rapidly up to a definite peak productive age, after which there is a gradual decline. Typically, productivity begins somewhere in the 20s of a scientist's life, attains the high point sometime in the late 30s or early 40s, and thereafter turns downward."[42] Or, as Einstein put it: "A person who has not made his great contribution to science before the age of thirty will never do so."[43] Harsh words, from our Einstein. Don't be discouraged though: you could always turn to philosophy, which knows no age limits.

The differences between the various scientific disciplines must also be taken into account: "In theoretical physics and pure mathematics both the ideation and the elaboration rates may be rapid, resulting in a curve that peaks comparatively soon and drops off much more quickly (...). In geology and applied mathematics, both the ideation and elaboration rates are probably slower, generating an age curve that peaks later and then declines more gradually."[44] To give some practical examples, let us cite, on the one hand, Arthur Cayley, a prolific British algebraic scholar of the nineteenth century, who published almost a thousand mathematical papers. On the other, the eighteenth-century British historian Edward Gibbon, who devoted practically his entire intellectual life to a single, very famous work, *The History of the Decline and Fall of the Roman Empire*. Or, if we want to remain in the field of the natural sciences, we shall name Copernicus, who devoted a good twenty years to the complex mathematical calculations needed to support

---

[41] Simonton, Dean K., Op. Cit., p. 59.
[42] Ibid. p. 66.
[43] Brodetsky, Selig, *Newton: Scientist and Man*, Nature 150, 698–699, 1942.
[44] Simonton, Dean K., Op. Cit., p. 72.

the validity of his *De revolutionibus orbium coelestium*. Still, on the subject of productivity, let us not forget the sometimes painful awareness of many scientists regarding the—inevitable?—decline[?] of their creative potential: for example, a study based on interviews with numerous academics from various disciplines revealed that only 33% of the physicists interviewed—with an average age of 47—believed they would be destined for further important research in the future, and only 17% of the mathematicians—with an average age of 53.[45] Don't lose heart, however: as you can see, these percentages are low, but not zero.

At this point, it is time to address the question with which we opened this paragraph, namely, that of the relationship between quality and quantity. Which is one of Simonton's fixed points: according to his Darwinian theory, the mind is, in fact, a producer of random mental combinations, so that, in order to arrive at quality, one must necessarily go through quantity. The scholar also calls into question the Scottish philosopher Alexander Bain, according to whom "the greatest practical inventions being so much dependent upon chance, the only hope to success is to multiply the chances by multiplying the experiments."[46] We could give many examples; one could be Edison, a certified genius, who put together no less than 1093 patents during his prolific career—including not only the electric light bulb, but also the phonograph and the movie camera. But let's not forget his equally numerous failures: a flying machine that failed to fly, for example, or an electric car that didn't go very far—the fault of the batteries—or even a house made entirely of concrete, including the furniture. All failures that did *not* go down in history, but which nevertheless testify to Edison's enormous productivity, a true embodiment of the trial-and-error principle. Beyond the specific cases, Simonton's reasoning—i.e., the link between quality and quantity—is essentially based on a fairly objective criterion, i.e., the number of publications of each eminent scientist and the number of citations that at least some of these publications receive—data in hand.[47] Simonton admits, however, that the correlation between quality and quantity is far from perfect; and so, while many scholars lie on the continuum ranging from the 'prolific'—scientists who produce a great deal, and with great success—to the 'silent'—those who produce little to nothing—there are also several exceptions. We also have the 'perfectionists'—who work on very little research, but what they do work on

---

[45] Simon, Rita J., *The Work Habits of Eminent Scholars*. Sociology of Work and Occupations, 1(3), 327–335, 1974.

[46] Bain, Alexander, *The Senses and the Intellect*, University Publications of America, Washington 1977 (1855), p. 597.

[47] See, for example: Dennis, Wayne, *Bibliographies of eminent scientists*, Scientific Monthly, 79, 180–183, 1954.

is of high value and requires a lot of effort—and the 'mass producers'—who produce a lot of work, but of little value.[48] A corollary of this idea is that it is by no means true, as the myth goes, that geniuses are always right; on the contrary, failures and mistakes will often be at least as numerous as successes. There are examples of this even among the greatest of the great: Galileo and his obsession with exclusively circular orbits, to the point of denying the physicality of comets, believing them to be atmospheric effects of the aurorae borealis; or Einstein, with his insistence on a completely deterministic universe; or Darwin, with his theory of pangenesis—which explained heredity through 'gemmules,' small hypothetical particles emitted by the various organs of the body and directed towards the gonads, so as to influence the gametes.

6.5   Ortega, Yuasa, Planck

Three themes are connected to the question of scientific productivity, namely, what Simonton calls the "Ortega hypothesis," the "Yuasa phenomenon" and "Planck's principle." Ortega's hypothesis starts from a sentence by the famous Spanish philosopher José Ortega y Gasset, who wrote, in *The Revolt of the Masses*, that "it is necessary to insist upon this extraordinary but undeniable fact: experimental science has progressed thanks in great part to the work of astoundingly mediocre men, and even less than mediocre. That is to say, modern science, the root and symbol of our actual civilisation, finds a place for the intellectually commonplace man and allows him to work therein with success."[49] How much truth is there in this statement? The term 'Ortega hypothesis' was originally coined by American sociologists Stephen and Jonathan Cole, who also proceeded to test it empirically. Based on the number of publications and, above all, citations, the two came to the conclusion that Ortega's claim, however well-intentioned—i.e., he meant it to combat elitism—was fundamentally false: eminent scientists would not rely much on the research carried out by their less eminent colleagues. Let us now turn to the Yuasa phenomenon. According to the Japanese physicist and historian of science Mintomo Yuasa, the focus of scientific activity in the modern world—which, according to his definition, consists of 25 per cent of scientific discoveries at any given time—tends to shift from one country to another every 80–100 years. According to Yuasa's data, the centre of scientific

---

[48] Terminology introduced in: Cole, Stephen & Cole, Jonathan, *Social Stratification in Science*, University of Chicago Press, Chicago 1973.

[49] Ortega y Gasset, José, *The Revolt of the Masses*, Norton, New York 1957, pp. 110–111.

production would have shifted from Italy (1540–1610) to, in order, Great Britain (1660–1730), France (1770–1830), and Germany (1810–1920), and then on to the United States (1920–), the current holders of the title. And it cannot be ruled out that a further shift is about to occur, e.g., to China.[50] That great geniuses are not evenly distributed throughout history is a well-known fact; it is known, for instance, that they tend to gather in clusters linked to a certain civilisation and historical epoch. To be precise, this fluctuation—both in terms of quantity and quality of contributions—is not cyclical, but aperiodic: that is, there is no general rule that establishes when genius will flourish in a certain culture and when it will fade away, there is only a tendency towards alternating periods of flourishing and periods—often much longer—of scientific and cultural apathy. Moreover, the flowering of thought in a certain nation can be 'contagious,' i.e., it can influence what happens scientifically and culturally in neighbouring nations. The reasons for this lie in numerous factors, and, in general, we can say that the cultural-political-economic milieu of a certain nation can change, and become more suitable for the flowering of scientific research. To this, we must add that, for Yuasa, the blossoming would be linked to the average aging of the scientific community. That is, when the average age of the research population exceeds fifty, the primacy of the scientific centre under consideration would be destined to decline. In order to talk about the Planck principle, we need to revisit Simonton's distinction between intuitive genius and analytical genius. In fact, as a scientist progresses in their career, their mind tends to move from an intuitive state to an analytical one, i.e., according to the scholar, they tend to acquire a vision of things that is solid, articulate and well organised, but at the price of their ability to accept radical cognitive changes. Hence, Max Planck's famous saying that "a new scientific truth does not triumph by convincing its opponents and making them see the light, but rather because its opponents eventually die, and a new generation grows up that is familiar with it."[51] Here, too, there is no shortage of anecdotes collected by Simonton, from Liebig vehemently attacking the theory of fermentation of Pasteur, a man 20 years younger than himself, to Freud confessing, in his *Civilization and Its Discontents*, that the conceptions he had advanced in his youth now exerted a very tight control over his worldview. There is also no shortage of empirical

[50] Yuasa, Mitsutomo, *The shifting centre of scientific activity in the West*, Japanese Studies in the history of Science 1, pp. 57–75, 1962.
[51] Planck, Max, *Scientific Autobiography and Other Papers*, Philosophical Library, New York 1949, pp. 33–34.

research on the Planck principle: we cite, for example, a study that high-lighted how age played some role in the acceptance of Darwin's theories.[52] Simonton comforts us a little, however; according to the scholar, reality is much more complex than that and—we emphasise—it is only a probabilistic tendency, which admits exceptions. Just to add an anecdote, let us recall that the Scottish geologist Charles Lyell, twelve years older than Darwin, was still doing research when Darwin's theory was made public, and his age did not prevent him from embracing Darwinian thought, even wondering why the idea had not come to him.

6.6   The Dilemma of Multiples

The time has now come to address a rather thorny issue concerning scientific genius. As anyone working in sociology or anthropology knows, these fields tend to minimise the role of the individual, indeed, in some cases, to deny its existence—that is, to regard the subject as a by-product of impersonal social-cultural-historical forces, not truly autonomous. Now, it is certainly true that the cultural context plays a very important role in facilitating or hindering the flowering of scientific genius; for example, we know that scientific progress is favoured by peace and hindered by war,[53] and that a cultural context charac-terised by materialism and empiricism is more favourable to techno-scientific progress than one dominated by mysticism.[54] However, here, the discourse is different: socio-cultural determinism is, in fact, an approach that erases subjectivity from history, and explains what happens with super-individual concepts, such as that of the 'zeitgeist,' i.e., the 'spirit of the times.' In essence, if a certain discovery or invention was made in a certain era, this is because it was somehow already 'in the air.' If Einstein had not formu-lated the theory of special relativity, someone else would have done it; if Edison had not invented the light bulb, another inventor would have done it; and so on. It is clear that, in an interpretation like this, the role of indi-vidual genius is reduced to nothing. Indeed, the socio-cultural determinism approach denies that such a concept has any interpretative usefulness—it would just be a cultural construct dating back to Romanticism, and best

---

[52] Hull David L.; Tessner, Peter D.; Arthur M. Diamond, Arthur M., *Planck's Principle: Do younger scientists accept new scientific ideas with greater alacrity than older scientists?* Science, vol. 202, 1978, pp. 717–723.

[53] Simonton, Dean K., *Interdisciplinary and military determinants of scientific productivity: A cross-lagged correlation analysis*, Journal of Vocational Behavior, 9, 1976, 53–62.

[54] Simonton, Dean K., *Do Sorokin's data support his theory? A study of generational fluctuations in philosophical beliefs*, Journal for the Scientific Study of Religion, 15, 1976, 187–198.

disposed of. Moreover, the reality of multiples would mean that scientific creativity and artistic creativity would function differently: if Michelangelo, Mozart or Picasso had not existed, their works would not have been created by someone else in exactly the same way; if they had not existed, art history would simply have been different. What evidence does the school of cultural determinism bring to support its theory that everything can be explained by the zeitgeist? The main element would be the so-called 'multiples,' i.e., all those discoveries and inventions that were made simultaneously by two or more people, independently of each other, confirming precisely that they were 'in the air.'

There are literally hundreds of cases of this, including some that are known to practically everyone. Famous, for example, is the dispute between Newton and Leibniz over who arrived at infinitesimal calculus first: apparently, Leibniz published his results first, but Newton's supporters accused the German philosopher of having plagiarised the English physicist's as-yet-unpublished ideas; and it is now believed that the two thinkers arrived at infinitesimal calculus independently. Or think of the calculations of Neptune's orbit by John Couch Adams and Urbain Le Verrier. Or the discovery of oxygen—Joseph Priestley and Carl Wilhelm Scheele -, the work on the law of conservation of energy—Julius von Mayer, Hermann von Helmholtz and James Prescott Joule-, the patents for the telephone— filed on February 14, 1876, by Alexander Graham Bell and, only a few hours later, by Elisha Gray—and those for anaesthesia—William Morton and Crawford Long. The case of Lassell and the Bonds is also emblematic: in 1848, the English astronomer William Lassell discovered Hyperion, a moon of Saturn, on the same night as two American astronomers, William and George Bond. Two years later, the scientist observed a dark inner ring around Saturn—dubbed the "crepe ring"—only to read the next day in the newspaper that this discovery had been made and promptly reported to the press, also by the Bonds. And the famous cases did not end there.[55] Various researchers have tried to compile more or less exhaustive lists of multiples; the well-known sociologist of science Robert K. Merton counted 264,[56] and Simonton as many as 579.[57] On the surface, all this seems to carry water for the supporters of Sociocultural Determinism. For example, according to one of its advocates, Merton, inventions and discoveries would be inevitable

---

[55] Cf. Simonton, Dean K., *Greatness. Who makes history and why*, The Guilford Press, New York 1994, p. 116.

[56] Merton, Robert K., *Singletons and multiples in scientific discovery: A chapter in the sociology of science*, Proceedings of the American Philosophical Society, 105, pp. 470–486, 1961.

[57] Simonton, Dean K., *Multiple discovery and invention: Zeitgeist, genius or chance?* Journal of Personality and Social Psychology, 35, 1603–1616, 1979.

once the necessary knowledge has been accumulated, if a sufficient number of people are working on a certain problem, and if social demands and those within a certain scientific discipline exert the necessary pressure. Cultural anthropology also had a hand in this; in particular, the American scholar Alfred Kroeber was impressed by the simultaneous rediscovery in 1900 of the laws of heredity—originally discovered by Gregor Mendel and then relegated to oblivion—by Hugo de Vries, Carl Correns, and Erich Tschermak. It could not have been by chance, according to the anthropologist, it had to happen, and right then and there.[58] Better therefore to take a look at a few objections to the zeitgeist theory. To begin with: although the number of multiples does indeed seem rather large, a closer look reveals that this is not really the case. In particular, several alleged copies are such only because they have been placed in categories that are far too broad. Let us take the example of nuclear magnetic resonance, which was discovered in 1946 independently by two different research groups, that of Ed Purcell, Henry Torrey and Robert Pound (at Harvard)[59] and that of Felix Bloch, William Hansen, and Martin Packard (at Stanford).[60] In regard to this, Purcell had to say that, although scientists "have come to look at the two experiments as practically identical (...), when Hansen first showed up [at our lab] and started talking about it, it was about an hour before either of us understood how the other was trying to explain it."[61] Identical general concept, then, but completely different concrete approaches.

One approach proposed by Simonton to reduce the number of multiples is to consider the true author of a certain discovery to be only that scientist for whom the discovery represents a central message. And so, both Julius Lothar Meyer and Dmitri Ivanovich Mendeleev worked on the classification of elements according to atomic weight, but only the latter made the well-known periodic law the central message of his work, to the extent that Meyer recognised his colleague's priority. The details and experimental predictions were worked out first and foremost by Mendeleev, to whom the honour of discovery therefore goes.

The same approach should therefore be used with Charles Darwin and Alfred Russel Wallace; the former's evolutionary theory was much more

---

[58] Kroeber, Alfred, *The superorganic,* American Anthropologist, 19, pp. 163–214, 1917.

[59] Purcell, Ed M.; Torrey, Henry C., and Pound, Robert V., *Resonance Absorption by Nuclear Magnetic Moments in a Solid*, Phys. Rev. 69, 37—1 January 1946.

[60] Bloch, Felix; Hansen, William W.; Packard, Martin, *The Nuclear Induction Experiment,* Physical Review, 70 (7). 474–485, 1946.

[61] Quoted in: Simonton, Dean K., *Origins of Genius—Darwinian Perspectives on Creativity*, Oxford University Press, Oxford 1999, p. 177.

comprehensive than the latter's, applied to every possible sector of the biological world, and, above all, it was intent on going where the other did not
dare, namely, towards human mental capacities—which Wallace continued
to disregard, prisoner, as he was, of a spiritualist view.

Or, to return to Mendel: according to some historians of science,
the latter's work was primarily concerned with hybridisation, whereas de
Vries, Correns, and Tschermak considered what was a peripheral aspect of
Mendelian studies,[62] to the extent that there are even those who would like
to deny Mendel himself the authorship of Mendelian genetics.[63] Sometimes,
then, the lists of multiples include discoveries that are not truly independent, in as much as the scholars and inventors involved actually influenced
each other. One example is the steamship: Robert Fulton did not, in fact,
build it independently of the Frenchman Claude-François-Dorothée de Jouf-
froy—the Marquis d'Abbans -, but relied on the latter's earlier work, and also
witnessed the experiments of another of the claimants to the title of father of
the steamship, the Scotsman William Symington.

Or think of the long-running controversy over the general theory of relativity, which everyone attributes to Einstein, but which, until the late 1990s,
was at the centre of a lesser-known dispute over priority and plagiarism.
While Einstein was immediately granted the title of discoverer, another scientist, behind the scenes, disputed its authorship, attributing it to himself; we
are referring to David Hilbert, one of the greatest mathematicians of the
early twentieth century. The clash with Einstein became rather unpleasant,
with accusations of plagiarism being exchanged; Einstein claimed that Hilbert
had copied his ideas from some of his early papers, while the mathematician's followers—Hilbert himself having died in 1943—claimed that the
real plagiarist was Einstein. This lasted until 1997, when three historians of
science, Leo Corry, Jurgen Renn and John Stachel, discovered that Hilbert's
final paper contained an ingredient taken from Einstein's work.[64] More
specifically, Einstein delivered his final paper on November 25, 1915, while
Hilbert's was delivered on November 20 of the same year, thus earlier, but
Hilbert's paper was not actually published until March 31, 1916. The original version delivered by Hilbert of the general theory of relativity was not
covariant—covariance is a principle whereby the form of the physical laws in
question must be independent of the co-ordinate systems under consideration—as Einstein's was, whereas the mathematician's final version from 1916

---

[62] Brannigan, Augustine, *The reification of Mendel*, Social Studies of Science, 9, pp. 423–454, 1979.
[63] Olby, Robert C., *Mendel no Mendelian*, History of Science, 17, pp. 53–72, 1979.
[64] Corry, Leo; Renn, Jurgen; Stachel, John, *Belated Decision in the Hilbert-Einstein Priority Dispute*, *Science*, 278 (5341): 1270–1273, 14 November 1997.

was covariant. Hence, Einstein could not have copied this last, crucial step—that of covariance—from Hilbert; on the contrary, Hilbert would have had to have adjusted his own work from Einstein's now published work. No multiple in this case then, but just the influence of one scholar on the other.

In addition, it must be said that certain cases sold as multiples are literally centuries apart, such as the discovery of Eustace's Tubes—Alcmeon of Croton in the fifth century BC and Bartholomew Eustace in the sixteenth century AD—or the observation of the passage of Venus in front of the Sun by the Islamic scholar Al-Kindi in the ninth century and the Englishman Jeremiah Horrocks in 1639. The very fact that some discoveries are made, forgotten and then renewed plays against the theory of cultural determinism: why does this happen, given that, according to CD, every discovery would happen precisely because it has to happen? If it has to happen, it should not be possible for it to fall into oblivion and then be rediscovered, even a long time later.

7

How to Become Who You Are

7.1   The Growth Mindset

Genius is a source of envy, at least in minds hungry for stimulation and dreaming of reaching unexplored heights of creativity. If you have felt small in the face of the great geniuses of humanity, if you think you will never be able to rival them, or even come close to rivaling them, this chapter is dedicated to you. Prepare yourself for an injection of optimism, although such a drug should still be taken in moderation. We are not about to provide you with some secret recipe for becoming a genius quickly and easily, because—as you will have realised by now—there is no shortcut to excellence. What we will do instead is introduce you to the work of some scholars of human talent who deny that it is an innate endowment. On the contrary, according to scholars such as Carol Dweck and Anders Ericsson, human nature—at least as far as individual limits and capabilities are concerned—is malleable, i.e., excellence and genius are, at least in theory, more or less within the reach of all those who are willing to make the necessary sacrifices to achieve them.

So, let us begin with the—now very famous—work of Carol Dweck, and, in particular, the ideas contained in her bestseller *Mindset. The new psychology of success*.[1] The American psychologist does not go into the true nature of human abilities, i.e., she does not actually tell us who wins in the eternal struggle between nature and nurture. Rather, her work concerns the power

---

[1] Dweck, Carol S., *Mindset. The new psychology of success*, Random House, New York 2006.

© The Author(s), under exclusive license to Springer Nature
Switzerland AG 2023
R. Manzocco, *Genius*, Springer Praxis Books,
https://doi.org/10.1007/978-3-031-27092-5_7

exerted on our lives by our own beliefs, especially those concerning our abilities and limitations. The things we believe to be true influence our lives, sometimes very profoundly. And changing such beliefs, however difficult that action may be, may, in fact, change our lives. The assumption—not elaborated by Dweck—is that human capacities can be cultivated, as long as they are believed to be so modifiable. Here, the psychologist thus introduces the construct that has made her famous, that of 'mindset.' In practice, she would have identified two types of mindset in human beings, the 'fixed mindset' and the 'growth mindset'. As can be easily guessed, the former concerns the idea that human capabilities are fixed, that they cannot be changed. Possessing such a view of ourselves only leads us to constantly doubt our capabilities. If we think that they are given once and for all, this will lead us to constantly put ourselves to the test in order first and foremost to prove ourselves, and we will end up doubting them every time we meet someone who seems to be more capable than we are; the fixed mindset would lead us to believe that the people who are better than us are so for 'structural' reasons, and, as such, the gap between us and them would be unbridgeable. Dweck reminds us, however, that Binet himself was of the idea that intelligence could be developed, that is, that we could become smarter. If this is true, then, technically, we are allowed to adopt the opposite mindset, that of growth. A mindset whereby the gifts we possess are developable beyond the present limits. Those who possess such a mindset—Dweck's research reveals—are more oriented towards learning and self-improvement, not to mention the ability to more realistically assess their own merits and weaknesses. And, just to name a few figures that Dweck labels as lacking brilliance in their youth, let us recall: Charles Darwin, Ray Charles, Marcel Proust, Elvis Presley and Jackson Pollock. So, changing the mindset can change the life. Agreed. But how does one do it? According to Dweck, it is a journey, i.e., a constant work of self-analysis, of paying attention to the explanations we come up with regarding our successes and failures, as well as to our inner monologue. Without forgetting that the scholar's model is one with clear boundaries, and that, in a real context, many people will present a mixture of the two mindsets—i.e., sometimes we explain what happens to us with one mindset, sometimes with its opposite.

7.2   Practicing as Hard as You Can

Let us now turn to the second of our scholars, and ask: what if it is not true that genius and human performance in general are the result of nature? Indeed, what if it is not even true that they are the fruit of mere cultural context? What if everything was instead down to the individual and their actions, day after day, moment after moment? What if, in short, there was a way to increase one's capabilities and performance, to boldly go—as the Star Trek tagline goes—"where no one has gone before?" What if, in other words, natural talent did not exist, and what if performance was the result of hard work and, above all, a specific method? Supporting this thesis—indeed, dedicating decades of research to it—was a Swedish psychologist, Anders Ericsson.[2] It is not biology, in its various incarnations—physiology, neuroanatomy, and whatnot—that produces talent. Yes, of course, there is a secret, but it is not in the genes, and it is something that everyone, at least in theory, can draw on. The source of genius and performance is, in fact, in what Ericsson calls "deliberate practice."

Let us return, then, once again to Wolfgang Amadeus Mozart, who—as we will see in a moment—is an ideal model for testing Ericsson's theory of deliberate practice. Those professionally involved in music will no doubt have heard of the so-called 'absolute ear,' also known as 'perfect pitch,' a term used to denote the rather rare ability to identify the absolute pitch—in other words, the frequency—of a musical note, without the aid of a reference sound. Besides being apparently a child prodigy, able to tour the major capitals and cultural centres of Europe from a very young age and amaze everyone with a variety of musical instruments, Wolfgang Amadeus Mozart apparently possessed the absolute ear. Even among great musicians, it seems that it is not particularly widespread. For Ericsson, this is a perfect example of a talent commonly regarded as innate, one that cannot be learned, and that instead falls into the territory of deliberate practice. The evidence for this is not lacking, starting with the fact that possessors of the absolute ear have apparently often received musical training at an early age, and that it is more widespread among populations speaking a tonal language—such as Mandarin Chinese and other Asian languages, in which the meaning of a certain word depends on the tone in which it is pronounced. Intent on studying the nature of the absolute ear, Japanese psychologist Ayako Sakakibara performed an experiment on twenty-four children, subjecting them to specific training aimed at teaching them to recognise musical notes without a reference sound. Although absolute ear affects, on average, one in every ten thousand people, all of the children in question managed to acquire absolute ear without any problems. The moral of the story: what was thought to be a natural talent turned out to be the result of a learning process focused on

---

[2] Ericsson, K. Anders; Pool, Robert, *Peak: secrets from the new science of expertise*, Houghton Mifflin Harcourt Publishing Company, New York 2016.

a specific goal.[3] Let's go back to Mozart: his father Leopold, an unsuccessful composer and violinist of limited talent, decided to turn his sons into top musicians. Projecting much, Leopold? Our story begins with Wolfgang's older sister, Maria Anna—aka Nannerl—who, at just over ten years old, played the harpsichord and piano to perfection. And with little Wolfgang, Leopold started working when his son was at an even earlier age. A mediocre musician perhaps, but an excellent teacher, Leopold, so much so that he wrote the first book dedicated to the pedagogy of music. By the time he had turned four, Wolfgang was practising full time—harpsichord, violin, and so on. Is it any wonder that, due to such—manic—training, little Wolfgang managed to develop an absolute ear?

In order to explain this phenomenon—a seemingly innate talent that is instead acquired through exercise—Ericsson calls into question neuroscience, and, in particular, the now well-known concept of neuroplasticity, i.e., the ability of our brain's networks of neurons to grow and change following some form of external stimulation. If, until the 1990s, scientists believed that the neural circuits of human beings—or rather, of all living beings, especially adults—were unchangeable, it now seems to have been established that they can indeed be reconfigured, also and above all as a consequence of exercise of any kind. Thus: for the classical version, talent—no matter in what field—is already inherent in the power of our nervous system, and exercise merely 'draws it out'; for the neuroplastic approach, talent is literally constructed through exercise, which modifies our brain. No predefined skills, then, but only those actually built through work and motivation. Of course, as far as learning and deliberate practice are concerned, childhood takes precedence over adulthood—for example, if you don't acquire the absolute ear by the age of six, then it becomes extremely difficult to do so, almost impossible, in fact, although there are exceptions. The primacy enjoyed by youth, however, should not discourage you if you are no longer very young: for Ericsson, organising proper training allows anyone to improve their skills, even beyond personal expectations.

So, is Ericsson's message that you have to work hard, full stop? Not quite. Mere hard work, according to the scholar, will get us nowhere. What makes the difference is, in fact, the approach, that is, the deliberate practice, which we are going to look at in detail in this chapter.

---

[3] Sakakibara, Ayako, *A longitudinal study of the process of acquiring absolute pitch: A practical report of training with the "chord identification method,'* in: Psychology of Music, Volume 42 Issue 1, January 2014.

7.3   The Principles of Deliberate Practice

Basically, for Ericsson, there would be a set of general principles that could be applied in any area of life, and that would enable—for those who apply them with dedication—very high levels of performance. And when we say 'any,' we mean 'any': from chess to music, from sport to management, from natural sciences to philosophy. With the understanding that what matters is the improvement of what lays at the root of performance, namely, our mental processes. In essence—even if Ericsson does not put it quite like this—his method is actually a meta-method, a set of recommendations that allow those who apply them to construct a method of action suited to the specific discipline they intend to practise. It also happens that there are disciplines—such as music or chess—in which teaching and process improvement already have standardised procedures for building expertise, whereas other areas of life—such as management or philosophy—do not. The principles elucidated by Ericsson are general, however, hence transdisciplinary, valid for every human endeavour. In essence, they are principles that can be declined according to this or that need, in order to construct procedures and routines perfectly suited to the discipline in which we wish to excel.

What Ericsson calls 'deliberate practice' or 'deliberate practice' [These are the same.] is, in fact, quite different from so-called 'naive practice,' which—according to the scholar's definition—consists in repeatedly performing a certain task or gesture, hoping that it will lead to improvement.

Basically, when we play a sport, or play a musical instrument, or do math, or whatever, and all we do is repeat the same routines automatically, well, that's 'naive practice'.

Deliberate practice consists in metaphorically injecting a great deal of thought into routine activities, setting and energetically pursuing very specific and increasingly ambitious goals that follow one another in an orderly fashion. It is therefore necessary to break more complex goals into smaller ones, and to establish a specific plan to achieve each of them. Deliberate practice is extremely, obsessively focused and fully self-aware.

Another aspect of Ericsson's methodology is the presence of a constant feedback process, possibly from third parties. And then, there is the issue of the comfort zone: it is necessary to get out of it systematically. In other words, deliberate practice cannot be pleasant, like a pastime is pleasant, but must be aimed precisely at overcoming the limits of the comfort zone.

Ericsson mentions our beloved Ben Franklin in this regard, particularly in reference to his attempt—unsuccessful it seems—to become a chess champion. One of the first chess players in America, he participated in the first championships held in the New World, devoting himself with increasing dedication to the game in question for a good fifty years. Despite this, Franklin made no particular progress, and found himself at a standstill, without even understanding why.

Today, we have the answers, at least according to Ericsson: Franklin would never have pushed himself outside the comfort zone, would never have injected massive doses of thought and effort into the game, instead continuing to play the same games as a hobbyist would.

Getting out of the comfort zone implies expressly trying to do something that you have never been able to do up to that point. However, it is not a matter of 'trying harder,' but of 'trying differently.' Of working out a personalised strategy to get around the block. And then continuing, until another blockage arises, for which a new strategy will be required.

Easier said than done, to be honest. But there is no shortcut to excellence, and, at this point, innate talent—which, for Ericsson, does not exist—gives way to a different factor: motivation. Specifically, what drives great performers?

In principle, motivation is maintained through positive feedback, be it internal or external, i.e., the pleasure we can take in seeing ourselves improve, or the approval that comes from others. Another possible motivating factor is a love of challenges: those who like to challenge themselves will tend to be much more motivated than those who are easily satisfied.

So: step out of your comfort zone, in a focused manner, with clear goals, and above all, devise a way—whatever it may be—to maintain motivation, which is the most important ingredient, even of supposed talent. Make no mistake, though: this seemingly simple recipe is only the beginning. The ingredients we have seen so far constitute a form of deliberate, purposeful, motivated practice, which, however, is not enough. Something else is needed, namely, Ericsson's deliberate practice. But before addressing this, we must resolve the issue of feedback, which, in the case of physical activities, is visible, while, in the case of mental activities, it is not. And to unravel this knot, we need to take a look at a very famous study, which you will find cited in many books on performance and which concerns a very specific group of professionals: London taxi drivers. We have already mentioned the concept of neuroplasticity, and now we need to take a closer look at this concept. We now know that the very functioning and structure of the brain changes with mental training, just as the body does—for example, the cardiovascular or muscular system—as a result of exercise. We know this because, for decades, we have had systems such as MRI—magnetic resonance imaging, a system that uses a strong magnetic field and radio waves to interact with hydrogen atoms in the brain and generate an image of its structure and functioning. In particular, for the subject of our interest, MRI-based studies have allowed us to verify the effects exerted on the brain by intense and repetitive activities such as those involved in deliberate practice. And so it is that our London taxi drivers come into the picture.

7.4   London Cabbies and the Brain

It's not easy to become a taxi driver in London. To do so, you have to memorise a complex map of all the streets, lanes and landmarks of the British capital, commonly known as "The Knowledge." How difficult could this be, you might say: quite difficult, actually, given that there are some twenty-five thousand streets and thousands of landmarks, including clubs, theatres, and whatnot. Three to four years of study, with a final exam that some have to attempt as many as a dozen times and that half of those who try ultimately end up flunking. What's more, according to a study published in 2011 in Current Biology, acquiring The Knowledge would cause structural changes in the brains of taxi drivers and, in particular, promote the development of the hippocampus, an area responsible for, among other things, the consolidation of memories and their transition from short-term to long-term memory, as well as spatial memory. The research was carried out by Eleanor Maguire and Katherine Woollett, two researchers from the Neuroimaging Centre at University College London; the scientists followed a group of 79 would-be taxi drivers, plus a control group of 31 people. In the course of the study, Maguire and Woollett analysed the volunteers' brains with MRI and also subjected them to mnemonic tests. At the beginning of the study, the hippocampi of the volunteers and the control group showed no difference. Over the course of the years—this was, in fact, a long-term study—39 would-be taxi drivers passed the difficult test, at which point the researchers had the opportunity to further divide the volunteers into those who passed and those who failed—plus the control group. The final result: those who passed the test experienced an increase in grey matter in the hippocampus, while those who failed and those in the control group experienced no change. However, this superiority only specifically concerns The Knowledge. For Maguire, his study represents confirmation that the human brain remains 'plastic' even in adulthood. Now, one could also argue that those who underwent an increase in the hippocampus were able to achieve such a result because the genes had already predisposed them to this possibility, i.e., that the potential for increase was already 'set' by the DNA; we therefore return to the usual Galtonian debate, 'nature versus nurture,' although, obviously, our Ericsson—as you will have understood—sides with nurture. A further note: according to the study in question, the taxi drivers who passed the exam performed worse than the failures and the control group in the use of visual memory for matters unrelated to The Knowledge and short-term memory. The moral of the story: according to the British researchers, there is only so much 'space' in the brain; if you occupy it with one skill, you cannot occupy it with other things. On this last aspect, we suspend judgement; future studies will tell us whether the human brain has 'territorial' limits or whether it is instead possible to develop one's mental abilities without losing ground in other areas of life. For the time being, we limit ourselves to recognising that this research represents perhaps the most striking proof of brain plasticity and of the fact that there may well be limits, although it is likely that they have not yet been reached—Ericsson's words.

But why is the human brain made this way? Why is it so different from a computer, which certainly does not have this ability to change its structure and functioning as a result of environmental stimuli? The origin of all this, the Swedish scholar tells us, is, paradoxically, a typical characteristic of living forms, homeostasis, i.e., the tendency to maintain equilibrium, i.e., in a certain sense, the status quo. Our bodies want stability. For example, body temperature is kept stable, although let us remember that this is a dynamic equilibrium, characterised by continuous 'adjustments' whereby we adapt to changes in the external environment while cultivating internal stability. However, this process of homeostasis has a side effect: after the compensatory adaptation has occurred, the organism can handle the external challenge more easily. For example, if we enjoy running and run often, our heart and lungs will adapt, and running itself will become easier. And, if we continue to push our limits, we will gradually find ourselves further and further removed from them. Let us mention at this point, along with Ericsson, a lipid-rich material that surrounds nerve fibres and allows for faster transit of electrical signals: myelin. Let us omit the technical details and simply observe that repetitive actions, such as practising sport or music, cause a 'myelinisation' of the nerve areas involved, so that—if one does not overdo it, i.e., if one does not push one's practice too far beyond the present limits—this process will result in a structural change, with the acquisition of new abilities. There is therefore a 'sweet spot,' a 'place' beyond one's limits, although not too much, and if one reaches it, one will be able to improve one's skills. And this applies not only to physical practices, but also to purely intellectual ones, such as mathematics or physics. Studies carried out on mathematicians and non-mathematicians alike[4] have shown that not only do those who have dedicated themselves professionally to mathematics have more grey matter in the areas deputed to these skills, but that this thickening depends on the length of practice: that is, the longer one has dedicated oneself to mathematics, the thicker these areas are, indicating that it is the practice itself that causes this change, i.e., that it is not the result of some genetic predisposition.

One thing must be kept in mind: acquired skills require maintenance. Stop training, and body and brain will slowly but surely return to where they started, a conclusion also confirmed by Maguire's study, whose taxi drivers, once they stopped working, witnessed a progressive reduction in the size of the hippocampus.

7.5   Chunks and Other Principles

At this point, Ericsson introduces another concept, which occupies a central role in his system of thought: that of "mental representation." Starting with the world of chess, whose champions famously manage to 'see' well in advance the possible scenarios that are about to unfold in the course of the game they are playing. The well-known American scholars Herbert Simon and William Chase, in an equally well-known study,[5] highlighted how good chess players manage to build patterns in their minds, through practice, that 'summarise' possible patterns of action on the chessboard, and that they call 'chunks.' In the minds of great masters, the number of these mental representations that summarise reality—enabling them to metaphorically see the forest while others see only separate individual trees—reaches, according to Simon, the dizzying figure of fifty thousand. Fifty thousand chunks that offer the player in question an overview that allows them to act with confidence and speed in a situation that the amateur would find tangled. Not only that, but the chunks would be organised hierarchically, from the most concrete to the most abstract. In other words, thanks to mental representations, masters—of whatever discipline—literally have the ability to 'see' the world, or at least certain sections of it, differently from us mere mortals. Obviously, the masters do not just see the forest: the skills they have acquired allow them to move up

---

[4] For example: Popescu, Tudor, et alii, *The brain-structural correlates of mathematical expertise*, Cortex 114 (2019) pages 140–150.

[5] Simon, Herbert A., and Chase, William G., *Skill in Chess: Experiments with chess-playing tasks and computer simulation of skilled performance throw light on some human perceptual and memory processes*, American Scientist, Vol. 61, No. 4, July–August 1973, pp. 394–403.

and down this hierarchy of chunks, quickly moving on to individual trees if necessary, focusing on the particular rather than the general.

But don't think that individual performances are the exclusive preserve of the great masters: we are all constantly building these chunks as we learn a new task. Think, for example, of the way we learn to ride a bicycle, or drive a car. Which allows Ericsson to add another detail: training based on mental representations is always specific. There is no such thing as generic ability. As if to say: if I train myself to play chess, I will become very good at chess, this skill cannot be transferred by weight to draughts or other board games.

Let us now look in more detail at the principles of deliberate practice. Let us first consider a fact, namely, that there are disciplines that lend themselves more to deliberate practice and others less so. It is a question of organisation, in the sense that disciplines such as chess, dance, or mathematics have—in some cases, for centuries—well-established and precise didactic methods, which allow those who try their hand at them to learn in a systematic way, reaching the level of master with due commitment. These 'fortunate' areas of life have certain characteristics in common. First of all, they have objective methods for evaluating performance. In the case of chess, the criteria are super-objective: the winner is better than the loser. Or of semi-objective methods, such as the judgement of expert judges. This is fundamental: if external judgments are missing, how can the level of a certain performance be assessed? Secondly, there is a certain degree of competitiveness in these disciplines, which drives their practitioners to develop increasingly sophisticated training systems. It is a virtuous circle: skills and training systems influence and enhance each other, following a trial-and-error procedure.

Other characteristics highlighted by Ericsson are solitary practice—training often consists of activities to be performed alone, even for activities that are later performed in groups, such as concerts and team sports—and the absence of fun. That's right: if you are engaging in deliberate practice and having fun, there is probably something wrong. This is because deliberate practice is about getting out of the comfort zone, and consequently cannot, well, be comfortable.

Here they are, then, the specific principles of deliberate practice. To begin with, if the discipline you have chosen belongs to the lucky ones, all you have to do is find yourself a good teacher. Indeed, deliberate practice requires precisely that the field in question be developed in this sense, that is, that it possess precise training systems and objective criteria of judgement and instructors with a sophisticated didactic method. More specifically, deliberate practice allows for the development of skills that others have previously learned to develop through a trial-and-error method. It is tedious and includes a very specific development plan—that is, it does not aim for generic improvement. A high degree of concentration and conscious action is also required of the practitioner. Feedback, then, is essential; it must be constant and specific. In time, the practitioner can learn to monitor his or herself, thanks also to the development of precise mental representations.

And what if the field we have chosen does not have established procedures, sound teaching and a competitive nature? In this case, the solution put forward by Ericsson is simple: apply the principles of deliberate practice as much as you can to your preferred field, inventing your own learning procedures and judging criteria, and perhaps competing with yourself or others. Easier said than done, you would say, and you would be completely right. After all, someone has to start with the trial-and-error labour. Ericsson's advice therefore is: get as close as possible, with ingenuity and creativity, to the principles of deliberate practice. If you lack masters and teaching methods, study the best performers, and try as much as possible to steal their secrets, to understand what differentiates them from other, lesser performers. Of course, sometimes there is a lack of criteria to establish who is really a top performer, in which case you will have to rely on a trial-and-error process. Once you have identified the expert and understood their method, copy it. If it works for you keep it; if it does not work drop it and move on.

7.6   Ten Thousand Hours?

And here, we finally come to the highlight of the chapter, the very famous—infamous?—ten-thousand-hour rule. In 1993, Ericsson, Ralf Krampe and Clemens Tesch-Römer published a famous paper,[6] later taken up by a well-known Canadian journalist, Malcolm Gladwell, in a bestseller published in 2008, *Outliers*, dedicated precisely to the theme of performance and excellence. And it was Gladwell who popularized among the general public the now well-known ten-thousand-hour rule, according to which ten-thousand hours of constant practice would make it possible to master any skill—a rule apparently based on the study of Ericsson, Krampe and Tesch-Römer. Easy to remember, the rule in question is not without charm in its simplicity. But what does Ericsson have to complain about? Many things, actually. To begin with, the 'rule'—let us use inverted commas at this point—is not so rigid; it depends on the sector that we consider. Some sectors certainly need ten thousand hours of deliberate practice, others more, others less—even much less.

---

[6] Ericsson, Anders K.; Krampe, Ralf Th., and Tesch-Romer, Clemens, *The Role of Deliberate Practice in the Acquisition of Expert Performance*, Psychological Review 1993, Vol. 100. No. 3, pp. 363–406.

Secondly, for each discipline, the rule in question represents a simple average of the hours dedicated to excelling; there are those among the top performers who needed more hours, and those who needed less. Third, Gladwell did not distinguish between deliberate and general practice. For example, in *Outliers*, he calculated the hours generically devoted by the Beatles to music, including concerts, wrongly presenting them as deliberate practice. Not to mention the fact that the Beatles' success was not due to their technical skills, but to the songs they composed, and therefore the time and energy devoted—particularly by John Lennon and Paul McCartney—to composition should have been analysed. Finally, Ericsson is keen to emphasise that it is not necessarily the case that a person who devotes ten thousand hours to practice will excel; his is not a rule, but merely an observation that there are no shortcuts to excellent performance. In fact, it requires an enormous effort, and the results are not necessarily guaranteed. The question as to whether the 'top' tier of performance is truly open to everyone still remains. What is certain is that, if one devotes ten thousand hours to deliberate practice, some degree of dramatic improvement will certainly be seen. But why then limit oneself to ten thousand hours? If one continues beyond this limit, one will certainly see further improvement, in what appears to be an upward spiral. Basically, one can never tell how far a properly motivated person will go.

7.7   You're Not Having Fun, Are You?

Remember: if you are having fun or if your mind is wandering pleasantly, you are probably not improving. Even in sports in which the mind is apparently not involved in any way, such as bodybuilding, concentration on the actions you are performing is of paramount importance.

And if you cannot find a suitable mentor or teacher? The example to follow is always our beloved Benjamin Franklin, who—in order to become a good writer—came up with a series of personalised techniques suited to the task. In particular, he regularly selected several articles he considered to be of quality, jotted down their contents briefly, then let his notes 'marinate' for several days, after which he tried to reproduce the original concepts in full—a way of 'forcing' his own style to improve. Realising the limitations of his vocabulary, Franklin also forced himself to use synonyms that he would not normally have thought of. By using techniques of this kind, the American genius was thus able to push himself beyond his comfort zone, demonstrating that he had intuited the principles of deliberate practice well in advance. Here again, the moral is always the same: if you don't find a good teacher or mentor, you will have to engineer and develop your own techniques for improvement. Which confirms our initial consideration, namely, that the one proposed by Ericsson is, rather than a model, a meta-model.

7.8   Deliberate Practice and Motivation

And if you reach a plateau, what do you do? If you are practising a skill, it will probably consist of several elements, and only some of them will block us, not all of them. So, for Ericsson, the solution will be to push the limits harder than usual, but only by a little. In this way, the incriminating elements will reveal themselves to us. And so, if you play tennis, try to challenge a better opponent than usual; if you are a manager, pay close attention to what goes wrong at the most difficult times; and so on. This attitude will reveal to you exactly what is holding you back, and allow you to then develop a specific plan to improve.

And for motivation, what do you do? Deliberate practice consists mainly of hard work, and maintaining motivation for such work over a very prolonged period of time, even several years, requires superhuman motivation, bordering on the manic. First of all, according to Ericsson, there is no such thing as generic willpower; rather, willpower depends on the situation. A person who shows great willpower in a certain sport, for example, might lack the same determination in another area of life. Moreover, so-called generic willpower is all too similar—for Ericsson—to the concept of talent, i.e., both are attributed to a certain person after events, i.e., after he or she has demonstrated a certain ability. In both cases, these are concepts that do not explain anything, because they are devoid of anticipatory power, i.e., they do not allow us to distinguish in advance who is gifted and who is not. They are 'circular' concepts: to devote oneself for years to the practice of music, for example, one must have great willpower. And how do I know that a certain person possesses great willpower? Because I have seen her practising for years! Better, then, to refer to motivation, which is a different concept from willpower. Motivation depends on two factors, namely, the reasons that drive us to continue and the reasons that drive us to stop. The secret, which is not really a secret, is to strengthen the former and try to weaken the latter. More specifically, we must try to creatively minimise—i.e., through development of a specific and personalised plan—everything that may interfere with our deliberate practice. Sooner or later—and this is the key to your motivation—what you do—sport, music, or whatever—will become an integral part of your identity. If a certain activity constitutes your identity—that is, if you feel, first and foremost, that you are a violinist, a tennis player, an intellectual, or whatever—your motivation will be strengthened. Underneath this 'identity,' motivation is the idea—also motivating—that a certain goal is within your reach, that you can achieve it. Everything that we have just elaborated about motivation takes us back once again to—guess who?—Benjamin Franklin. As a young man in Philadelphia, he created a club ('the Junto') whose social purpose was the mutual improvement of its members, especially from an intellectual point of view. The rules of the club prohibited members from contradicting each other or expressing their opinions too strongly. Every member was required to ask at least one interesting question every week, which forced all members to rack their brains. Don't underestimate Franklin's idea: making intelligent use of the pressure normally exerted by one's peers can boost our motivation dramatically. Without forgetting that, by its very nature, deliberate practice is a solitary activity, and therefore motivation requires additional techniques. For example, one can break a skill into more circumscribed skills, so as to work on them and see improvements day by day, and thus keep oneself motivated.

So, to sum up: according to Ericsson, by systematically applying the principles of deliberate practice, anyone—but especially the youngest—could, at least theoretically, achieve performances equal to those of humanity's greatest geniuses. Now, a question arises. Is this really possible? Has anyone tried to breed a genius on purpose? Is there, that is, an experiment, an empirical proof that shows that anyone can potentially become a genius? A rhetorical question of course, one that, to answer, we turn to a well-known Hungarian scholar, László Polgár, and his—interesting? Original? Strange?—family.

7.9   The Extraordinary Polgár Sisters

Born in 1946, Polgár was a psychologist and chess teacher, father of the even more famous Polgár sisters, whom we will come to in a moment. Having been interested in the subject of intelligence and genius since his university days, Polgár had been concerned with the lives of geniuses, noting that, at least in many cases, they would have begun to devote themselves to a certain discipline at a very young age and then studied it intensively. Having studied the biographies of some four hundred 'greats,' including Einstein, Polgár came to the conclusion that, if one takes the right approach, it is possible to take any child and make him or her reach heights of performance and genius. If a child is born healthy, says the scholar, they are already potentially a genius. And, in order to prove this, Polgár began a singular experiment. In 1965, the psychologist wooed—by letter—a Ukrainian foreign language teacher, Klara, to whom he outlined his plan: to marry and have children to be transformed, through the educational process, into geniuses.

The couple gave birth to three daughters, Zsuzsa (1969), Zsófia (1974), and Judit (1976), who were educated at home. Their parents taught them various subjects: mathematics, foreign languages, but especially chess. The choice fell on this game because of its measurability. The excellence of a chess champion is, in fact measured, objectively, through their victories and defeats. The experiment officially began in the early 1970s. Polgár began teaching the game of chess to the first of his daughters, Zsuzsa, when the latter was four years old.

All three sisters became extraordinary chess players. Zsófia, who became the sixth best female chess player in the world, stopped playing and decided to devote herself to something else—in particular, painting and interior design. Judit became the number one female chess player, followed in second place by her sister Zsuzsa. Notably, at the age of four, Zsuzsa overwhelmingly won her first tournament, the Budapest Girls' Under-11 Championship; about ten years later, at fourteen, she became the best female chess player in the world, and was later awarded the title—until then, reserved only for men— of "grandmaster." Judit—the greatest success of Polgár's experiment—became grandmaster at the age of fifteen, and thereafter world champion of the women's circuit for twenty-five years—until her retirement from the chess world in 2014. In 2005, she also became the first woman to participate in the World Chess Championship. For Polgár, genius would be the result of a simple formula, Labor plus Luck. Child prodigies would therefore not be miracles, but natural phenomena.

His experiment—educating children to become geniuses—has actually been conducted many times in history—think only of Mozart—and the only difference between his case and previous ones lies in the fact that Polgár's experiment was done with the conscious and intentional goal of producing a genius. Or rather, to grow a happy genius. The essence of Polgár's pedagogical system is that any healthy child can grow up to be a genius in a specific, specialised field. What is needed are parents who are 'obsessed' in a good way, says the psychologist.

According to the scholar, human personality is the result of three factors, namely, the natural endowment—genetics—the surrounding environment and individual creation. It is not possible to know with certainty what genius depends on, but, for him, the hypothesis that it is the environment—with the collaboration of individual motivation—that shapes it cannot be ruled out. The child must, in any case, be considered co-author of its own development. Every healthy person would have sufficient natural endowment at their disposal to become a genius. There are therefore no 'innate' abilities to be discovered, but rather abilities that can be built in everyone. Things, however, must not be left to chance.

In Polgár's pedagogical system, the period between the ages of three and six is central, and using this interval to cultivate a child's genius does not mean depriving them of their childhood, for, at that age, even efforts at self-improvement can take on the aspect of play, i.e., there is no real distinction between work and play, provided things are organised in the right way. The age at which one begins preparation is crucial, in the sense that, as time passes, the chances of becoming a genius decrease. It is almost impossible to become a genius by starting the preparatory work in adulthood, apart from certain fields, according to the Hungarian scholar, such as literature and philosophy, in which the doors of excellence are open even to people who are no longer very young. For Polgár, the secret of training for genius is the choice of a specialised activity and intensive instruction—his daughters, from the age of four, played chess five or six hours a day. In order to teach his daughters to play chess, Polgár broke up the game in question into simpler games, e.g., relating to recognising the board and using individual pieces, and then gradually added up the individual games until they were playing the complete game of chess.[7]

7.10 Deliberate Practice and Innovation

So, let us return to Ericsson, and ask ourselves: how do we deal with true creatives, those who do not merely follow existing paths, but discover new ones? How does creativity fit into the framework of deliberate practice? If such a procedure seems to be suited to the development of skills related to already established disciplines, how do we explain the mastery of people who invent completely new systems or disciplines? How do you explain an Einstein, a Darwin, a Beethoven or a van Gogh? All people who—mind you—moved from the known to the unknown, exploring new territories for which a methodology for excelloong was not even vaguely present. Ericsson's model also deals with this aspect, which represents the final level of his theory on expert performance. What is certain about innovators is the fact that, almost without exception, they—before innovating—worked for a long time

---

[7] Polgar, Laszlo, *Raise a Genius!*, Gordon Tisher 2017, Vancouver, https://slatestarcodex.com/Stuff/gen ius.pdf.

to become experts in their field. This makes sense: it goes without saying that, before contributing something new and original to physics or mathematics, one must master the discipline in question, or at least the specific field in which one wants to contribute. The same goes for any other discipline or practice, from athletics to piano, from painting to engineering.

And, speaking of painting, Ericsson tells us again, take the case of Pablo Picasso as an example. It is true that, looking at his works, especially his most famous ones, one might think that they sprang spontaneously from his brilliant mind, as they seem to break all ties with tradition. In reality, Picasso began by painting in a more or less classical style, over which he demonstrated mastery—and here, we are in the presence of classic deliberate practice. He then experimented with various artistic styles, modifying them until he reached the style that most distinguished him in the public perception. The point of the discourse Ericsson presents us with is that there would be no real difference between the deliberate practice followed to achieve mastery in an established discipline and the creative act that breaks new ground. In essence, as the scholar puts it, what expert performers do to move the boundaries of a discipline is not so different from what they do to reach the boundaries in question. After all, what great performers have acquired from their teachers is the ability to improve themselves, and they have become great performers precisely because they have not merely imitated their teachers. The metaphor Ericsson uses is that of a ladder, on which we climb step by step, dedicating our lives to building up each individual step, until we reach the highest level reached. From this point on, we can no longer be guided by teachers, and we may not know exactly where we are heading, but we more or less have an idea of what we need to do to add one more step to the ladder. Sometimes, pathbreakers intuit what they want to achieve, and their work then becomes a series of iterative operations in which they return to the problem over and over again, which is precisely the same process used to achieve mastery. Not even the 'aha' moments could exist without the slow and intense work that led to the construction of the ladder in question, a ladder to which a further piece is then added. Creativity would seem, for Ericsson, to be very similar to deliberate practice, in particular, its central ingredient, the need to maintain very intense concentration over a prolonged period of time. At this point, the Swedish scholar surrenders to the mystery: creativity, he writes, will always retain some unknown aspect, because it consists in producing precisely something that was not there before, but the concentration that characterises it will not be so different from that of deliberate practice.

7.11 Mozart and Other Prodigies

And now, it is time to apply Ericsson's theories in full to what, together with Franklin, we have considered in this chapter a real 'textbook case': Mozart. Indeed, the great musician is often cited as an example of a child prodigy endowed with a natural talent that cannot be explained by the theories of deliberate practice. As we have already mentioned, at the age of six, the young Wolfgang was taken by his father, together with his sister Maria Anna, on a tour around Europe, where the two youngsters were able to amaze listeners with their musical performances. Today, however, such performances seem much less miraculous, given that it is often the case that children as young as five or six can play the piano or violin at comparable levels, thanks to the Suzuki method.[8] Mozart obviously did not have the Suzuki method at his disposal, but, on his side, he had a father devoted to the idea of raising a musical prodigy and specialised in musical didactics—a discipline to which he even dedicated a book, testing its contents first on his daughter and then on Wolfgang, who began his journey very early, probably before the age of four. This therefore explains Mozart's musical talent. But what about his composing skills? Little Mozart, in fact, began composing music at the age of six and wrote an opera at twelve. It should be noted, however, that, unlike the Suzuki method, which only concerns musical performance, Leopold taught his son composition from an early age. We certainly know that little Wolfgang's earliest compositions are written in Leopold's handwriting, so we do not know for certain how much the latter 'cleaned up' his son's work. The first compositions directly attributable to Mozart date from when the musician was fifteen years old, i.e., a full decade after the young man began training to compose. Undoubtedly, therefore, this is a great genius of humankind, but—according to Ericsson—his compositional skills are far from inexplicable.

At this point, let us take advantage of the child prodigy Mozart to connect to the case of savants. How does the theory of deliberate practice explain the ability of some people with mental disabilities to perform, for example, very complex mathematical operations or other prodigious performances that are impossible even for ordinary people who do not have these disabilities? We are talking about a phenomenon known as the 'savant syndrome,' which concerns extraordinary abilities that arise in very specific areas. Some savants

---

[8] The Suzuki method consists of a pedagogical and musical technique developed by the Japanese violinist Shinichi Suzuki (1898–1998) and consists in establishing a formal parallel between the learning of a musical instrument and the linguistic environment in which children acquire their native language. In practice, it prescribes the musical saturation of the learner's environment.

are able to play a musical instrument and memorise thousands of different pieces, others are able to produce works of art, still others perform astounding mathematical calculations, or master the calendar in astonishing ways—for example, they can guess the day of the week of any given date, future or past.

Two scholars from King's College London, Francesca Happé and Pedro Vital, who compared autistic children with savant syndrome and autistic children without such abilities, were the ones to address this issue and, in particular, the nature of these extraordinary abilities.[9] The two scholars noted that the former were characterised by a tendency to produce repetitive behaviour in an obsessive manner, much more so than the latter. When something caught the savants' attention, they tended to focus on it, completely ignoring everything else. Here, then, is at least one of the savants' secrets, obsessive practice, which reminds us precisely of our beloved deliberate practice and the precisely obsessive commitment it requires.

Given the importance that deliberate practice has—indeed, always had—for human nature, Ericsson finally proposes changing the name of our species from 'Homo sapiens' to 'Homo exercens,' i.e., practicing man. What we are is thus, for the Swedish scholar, the fruit of our actions and focused attention. For human beings, to exist is to practise with method and dedication.

7.12 To the Master Goes the Knife

And now, to finish this chapter, we want to indulge in self-help literature for a moment, and take a look at a text that has done quite a bit of talking: *Mastery*, by Robert Greene, a true best-selling author who has dealt with seduction, power, strategy, and now also mastery.[10] Let's see, then, what Greene recommends in order to achieve excellence. To begin with, the author reassures us: it is never too late to start the process of developing mastery. There are also, according to him, rules regarding apprenticeship that transcend disciplines, time and space. In particular, one must be able to embrace tedium, which is a necessary stage in our journey towards excellence. We must then take great care in seeking out a master, in absorbing all we can from them, but avoiding remaining in their shadow: the goal must always be to surpass the master. The difficulty, of course, lies in finding the right mentor. Contrary to what is said about many geniuses, i.e., that they are self-centred, for Greene, the master or

---

[9] Happé Francesca, and Vital, Pedro, *What aspects of autism predispose to talent?* Philos Trans R Soc Lond B Biol Sci. 2009 May 27;364(1522):1369–75.

[10] Greene, Robert, *Mastery*, Viking Adult, New York 2012.

future master must know how to move effectively in the social environment, so as to be able to promote his or her work.

And what, for Greene, is mastery? A form of superior intelligence that is achieved by fully immersing oneself in a certain area of knowledge and respecting our natural inclinations. And to achieve it requires not ten thousand, but a good twenty thousand hours of practice: first ten thousand to master the art, and then another ten thousand to surpass it, i.e., to become truly original—it is not clear where this number of twenty thousand hours comes from, but we can live with it. Creativity, for the author, does not appear out of nowhere, and mastery would be the feeling of control over reality, or rather over a certain sector of reality, the sense of mastery. Greene divides this process into three phases: the apprenticeship, the active-creative phase, and mastery itself. In a sense, the master completes a cycle, starting from curiosity and returning to a childlike state of curiosity and direct interest in the world, but at a higher level. Thanks to mirror neurons—the author speculates— the master develops an ability to 'think within', to be able to identify with physical forces, for example—as Michael Faraday would have done—or with anatomy—in the case of Leonardo. Greene, at this point, compares Galton, with his exceptional IQ, and Darwin, who described himself as 'a very ordinary boy', but who evidently possessed something that Galton, who had a brilliant career, did not. A mediocre student, Darwin nevertheless followed his natural inclination, devoting energy and concentration to it and manifesting a desire to learn as much as possible in relation to his chosen field, with an obsession that dominated his daily thoughts. So, quite simply, the first thing you must do is to discover your vocation, dedicating yourself as much as possible to self-exploration: the first journey towards mastery is inward. By searching, sooner or later, we will end up finding our field of interest or, as they say in the US, our mission and vision, a purpose, according to Greene, that is bigger than ourselves. The worst thing a person can do is to pay attention to the crowd, instead of to one's inner calling, and stop trusting one's vocation. What should one do, then, to find one's calling? Greene offers us five strategies. To begin with, one can return to one's original inclinations, dating back to childhood, or stay with them—Greene cites the story of Einstein's fascination as a child with the compass given to him by his father. In short, to master a subject, one must love it passionately; for Einsten, it was not physics as a discipline, but, more generally, the deeper mystery of how the universe works. Second strategy: to occupy the perfect Darwinian niche: that is, to find an as-yet-unexplored territory in the working-professional-intellectual ecosystem. Third, the strategy of rebellion: just like Mozart, who, in 1781, decided to stay in Vienna and never return to Salzburg, rebelling against his father and beginning to compose furiously. Four, the strategy of adaptation: finding ways to adapt one's mission to the circumstances that fate has chosen for us. Five, the life-or-death strategy, that is, seeing the pursuit of one's mission as a matter of life or death, because, indeed, it is.

Having done so, one must submit to reality: that is, one must move on to the apprenticeship stage, just as Charles Darwin boarded the Beagle to embark on his crucial voyage around the world. More specifically, one must choose places and jobs that offer the greatest opportunities for learning. Apprenticeship, in turn, comprises three phases: the passive mode, consisting of deep observation, the practice mode, in which one acquires a certain skill, and the active mode, in which one experiments, without waiting until one feels ready. In one's confrontation with reality, one must give more weight to learning than to money, expand one's horizons and rely on one's sense of inferiority. One must also trust in the process of learning, and move towards endurance and pain. When he set out to write the poem 'Endymion,' John Keats gave himself an impossible deadline, seven months, with the goal of writing fifty lines a day: all this in order to avoid wasting attention. Life is short, one has to concentrate, and absorb the power of the master; Greene cites the case of Michael Faraday, who studied with the chemist Humphry Davy, but abandoned him when he felt ready to go his own way, in deference to the Spanish saying 'al maestro, cuchillada,' meaning, more or less—in a metaphorical sense—'one must stab the master.' There are, of course, exceptions here, too: this was the case with Thomas Alva Edison, who had to rely on books and ideal masters, because he did not have the opportunity to find himself a mentor, thus developing extreme self-reliance.

In terms of social intelligence, Greene recommends seeing people for what they are, without trying to change them. And then, there is faith, not in the religious sense, but in the sense of patience and trust that your work will bear fruit: there must, in short, be an obsessive element. This is the only way to achieve mastery.

8

Genius in the Machine

8.1 Machines that Create

Not content with studying human genius and unlocking its secrets, several scholars have sought to reproduce its capabilities in machines. This has led to the birth of a fascinating field of research—strictly interdisciplinary—known as Computational Creativity. In particular, the disciplines contributing to this endeavour—giving creativity and originality to machines—are artificial intelligence, psychology and, of course, the arts in general. More specifically, the aim of Computational Creativity is both to enable computers to achieve a degree of creativity similar to that of humans, to better understand—and model—human creativity, and to develop programmes that humans can use to enhance their own creative abilities. An ambitious project, as you can see, and not without theoretical and practical problems. And the first problem is precisely a logical one: if creativity consists in innovation—that is, in breaking established rules and replacing them with new ones—how can one think of developing an algorithm—that is, a set of rules to be followed during a calculation procedure—that is itself creative? This is a problem that was already pointed out decades ago by Teresa Amabile.[1] How, in other words, can a computer—a machine that merely executes what it is told to do—be creative? And already with this statement, we run into a problem: not all theorists of artificial intelligence agree with the idea that computers only do what they are programmed to do; in fact, as early as 1967, Marvin Minsky railed against

---

[1] Amabile, Teresa, *The social psychology of creativity*, Springer-Verlag, New York 1983.

© The Author(s), under exclusive license to Springer Nature Switzerland AG 2023
R. Manzocco, *Genius*, Springer Praxis Books, https://doi.org/10.1007/978-3-031-27092-5_8

this idea.[2] Since psychology itself does not offer an unambiguous definition of creativity, it is clear that theorists who want to develop a computational version of it will have to choose one of the available definitions and use it as a working tool. Three American scholars, in particular, were concerned with this issue between the 1950s and 1960s. Between 1955 and 1956, in Santa Monica, the three scientists—economist Herbert A. Simon, computer scientist Allen Newell and systems programmer John Clifford Shaw—developed the first programme aimed at imitating human problem-solving abilities. Subsequently, the scholars proposed a definition of creativity that characterises as creative any solution to a problem that is new and useful, that rejects previously accepted ideas, that is the result of intense and prolonged motivation and that makes a previously vague and obscure problem clearer.[3] The three scientists' approach is top-down, i.e., it starts with a definition of what human creativity is and tries to simulate it. This was complemented, during the 1980s and 1990s, by a bottom-up approach, which—starting with artificial neural networks, which simulate the basic structure of the brain—attempts to construct higher mental functions, including creativity, from the bottom up. One of the incarnations of this approach is genetic algorithms, procedures that simulate the process of natural selection to generate sets of possible solutions to a certain problem. The applications of genetic algorithms are numerous to date: they range from simulations in physics and chemistry, to the optimisation of molecular structures, the development of electronic circuits, the search for bugs in hardware, the development of models for the simulation of climate change, the development of advanced trading systems in the financial sector, architectural design, and much more. Traditional approaches are therefore based on explicit rules (top-down) and random variation (bottom-up); at this point, however, something new enters the scene: Machine Learning. This term refers to a field of research dedicated to the development of methods that make it possible to use progressively collected data in order to improve—just as progressively—the performance of a certain programme. More specifically, we are dealing with algorithms that build a model from a database of data—known as 'training data'—enabling it to make decisions without explicit request from the programmer. Just for the record, the e-mail service you use probably uses such algorithms when it decides that a certain e-mail should go to spam rather than to the in-box. And

---

[2] Minsky, Marvin, *Why programming is a good medium for expressing poorly understood and sloppily formulated ideas*, in: Krampen, Martin & Seitz, Peter (Editors), *Design and planning 2: Computers in design and communication*, Hastings House, New York 1967, pp. 120–125.

[3] Newell, Allen; Shaw, John Clifford, and Simon, Herbert A., *The process of creative thinking*, in: H. E. Gruber, G. Terrell and M. Wertheimer (Eds.), *Contemporary Approaches to Creative Thinking*, pp. 63–119, Atherton, New York 1963.

of course, there are those who have thought of using such a strategy to simulate human creativity.[4] In particular, the aim of this technique is to provide computers with starting data that enable the non-linear—hence 'emergent'—production of creative products. As early as 1989, an American scholar, Peter Todd, trained a neural network for the first time to generate certain melodies from a set of pieces of music used as training data; in this case, the cognitive psychologist used a specific change algorithm that allowed the computer to reproduce new music, without control, starting precisely from the music provided to it at the beginning.[5] Later, in 1992, Todd employed a different approach, known as the 'distal teacher approach,' based not on one, but on two neural networks, one of which provides the training data to the other.

Alongside experimentation and practical research, there was, of course, no lack of theoretical work. And so, the field of computational creativity has dabbled with a whole series of exquisitely philosophical questions. Let us therefore take up again the arguments that we have been making about creativity, quoting the cognitive psychologist Margaret Boden; the British scholar distinguishes between P-creativity—relating to the creation of something that is new only for the subject—and H-creativity—historical creativity, relating to the creation of something that is also new for a certain society.[6] A further distinction made by Boden is that between exploratory creativity—consisting of experimentation of the new within a specific conceptual space—and transformative creativity—more radical, consisting of the transcendence of the space in question. Let us not forget that Boden works in the field of artificial intelligence, and that her theoretical contributions serve as a framework for the development of algorithms; in fact, they have been formalised, in particular, by Geraint Wiggins, a British scholar of computational creativity.[7] In any case, these are very abstract models, from which one then tries to give direction to applied research. To give another example, let us cite the work of two American researchers, Mark Turner and Gilles Fauconnier, who propose a model christened Conceptual Integration Networks and based on Arthur Koestler's theory of bisociation.[8] The model in question includes four

[4] Mateja, Deborah; Heinzl, Armin, *Towards Machine Learning as an Enabler of Computational Creativity*. IEEE Transactions on Artificial Intelligence. 2 (6): 460–475, December 2021.

[5] Todd, P.M., *A connectionist approach to algorithmic composition*. Computer Music Journal. 13 (4): 27–43, 1989.

[6] Boden, Margaret, *The Creative Mind: Myths and Mechanisms,* London: Routledge, London 2004.

[7] Wiggins, Geraint, *A Preliminary Framework for Description, Analysis and Comparison of Creative Systems*, Journal of Knowledge Based Systems 19(7), pp. 449–458, 2006.

[8] Fauconnier, Gilles, Turner, Mark, *Conceptual Integration Networks*, Cognitive Science, 22(2), pp. 133–187, 2007.

connected conceptual spaces: a first input space, a second input space—to be mixed with the first—a generic space—which contains schemata and conventions that allow the first two spaces to be interpreted—and a blend space—a conceptual place where inputs from the first two spaces are mixed. In 2006, Francisco Câmara Pereira—a scholar specialising in Machine Learning—developed such a model, generating genetic algorithms capable of precisely blending different elements—linguistic or visual—thus enabling computers to produce, for example, three-dimensional images of mythological monsters.[9] Worthy of note is the attempt to attribute to machines the ability to invent stories. As early as the 1970s, the American scholar James Meehan developed a programme (code name: TALE-SPIN) capable of writing short stories structured as problem-solving efforts by characters.[10] Next, we mention BRUTUS, created by the US scholar Selmer Bringsjord and capable of developing more complex stories with complex interpersonal relationships.[11] Developed by Rafael Pérez y Pérez and Mike Sharples, the MEXICA model allows content to be generated without a predefined objective, but following rhetorical restrictions.[12] Another area of interest is undoubtedly that of computational humour, i.e., the attempt to develop programmes capable of inventing jokes and making jokes. This is an area that requires the management of complex information distributed on several levels—think of double entendres, puns, and so on. In regard to the field in question, we certainly must mention Kim Binsted and Graeme Ritchie, British artificial intelligence scholars, who specifically dealt with riddles[13]; the Dutchmen Hans Wim Tinholt and Anton Nijholt, who developed a programme that deliberately misinterprets the information provided[14]; and the work of Oliviero Stock and Carlo Strapparava, who developed a system for the production of humorous acronyms.[15]

---

[9] Pereira, Francisco Câmara, *Creativity and Artificial Intelligence: A Conceptual Blending Approach*, Applications of Cognitive Linguistics: Mouton de Gruyter, Amsterdam, 2006.

[10] Meehan, James, *TALE-SPIN*, in: Shank, R. C. and Riesbeck, C. K., (eds.), *Inside Computer Understanding: Five Programs plus Miniature, Hillsdale*, Lawrence Erlbaum Associates, Hillsdale 1981.

[11] Bringsjord, S., Ferrucci, D. A., *Artificial Intelligence and Literary Creativity. Inside the Mind of BRUTUS, a Storytelling Machine.*,: Lawrence Erlbaum Associates, Hillsdale, 2000.

[12] Pérez y Pérez, Rafael, Sharples, Mike, *MEXICA: A computer model of a cognitive account of creative writing*, Journal of Experimental and Theoretical Artificial Intelligence, 13, pp. 119–139, 2001.

[13] Binsted, K., Pain, H., and Ritchie, G., *Children's evaluation of computer-generated punning riddles*, Pragmatics & Cognition, 5 (2): 305–354, 1997.

[14] Tinholt, Hans Wim & Nijholt, Anton, *Computational Humour: Utilising Cross-Reference Ambiguity for Conversational Jokes*, In: Masulli, F., Mitra, S., Pasi, G. (eds) *Applications of Fuzzy Sets Theory. WILF 2007. Lecture Notes in Computer Science*, vol 4578. Springer, Berlin, Heidelberg, pp. 477–483.

[15] Stock, Oliviero; Strapparava, Carlo, *HAHAcronym: Humorous agents for humorous acronyms*, Humor: International Journal of Humour Research, 16(3) pp. 297–314, 2003.

And what would life be without a bit of poetry? This is what the various scholars who have tried to develop programmes capable of producing poetic compositions must have thought. For example, back in 2001, the Spaniard Pablo Gervás developed ASPERA, a programme capable of producing poetic verses from inputs consisting of extracts from existing poems.[16] And from poetry, we move on to music: there is no shortage of programmes capable of composing musical pieces, both classical—software capable of producing pieces in the style of Mozart or Bach—and modern. Let us not fail to mention the American David Cope, creator of a programme christened EMI ('Experiments in Musical Intelligence'), sophisticated enough to convince human listeners that they are dealing with compositions created by the musicians it imitates.[17] However, stylistic imitation is not enough: to be musically creative, a computer must also be able to compose original stuff. Here, then, is Iamus, a computer cluster located at the University of Malaga in which Melomics—which stands for 'genomics of melodies'—is embedded, a computational system whose algorithms allow for the 'evolutionary' generation of new works of contemporary classical music. If, on the other hand, you are more interested in jazz improvisation, then you should turn to Shimon, a robot developed by a Georgia Tech scholar, Gil Weinberg, capable of just such a performance.[18] And this is just a small taste: in recent years, several music-generating programmes have sprung up. Let us now turn to the skill that impresses the general public the most, namely, computational creativity in the visual arts. Investigating a little in this area, we soon discover that computers are now capable of producing both realistic and abstract works. In operation since 1973, AARON—developed by a British artist, Harold Cohen—has undergone a process of constant improvement over the years, to the point of producing works deemed worthy of being exhibited in respected galleries, such as the Tate Gallery in London. Then, there is NEvAr, a system (full name: Neuro-Evolutionary Art) developed by Portuguese scholar Penousal Machado that uses a genetic algorithm to create paintings in several stages—each stage is accompanied by human intervention, in the sense that a human operator chooses the best paintings developed, which then undergo further refinement until the final work.[19] Then, there is The Painting Fool, a system similar to AARON and developed by British

---

[16] Gervás, Pablo, *An expert system for the composition of formal Spanish poetry*, vol. 14, Journal of Knowledge-Based Systems, pp. 181–188, 2001.

[17] Cope, David, *Computer Models of Musical Creativity*, Cambridge, MIT Press, Cambridge 2006.

[18] *A Robot Named Shimon Wants To Jam With You*. NPR.org. 22 December 2009. https://www.npr.org/2009/12/22/121763193/a-robot-named-shimon-wants-to-jam-with-you.

[19] Machado, Penousal; Romero, Juan, eds., *The Art of Artificial Evolution: A Handbook on Evolutionary Art and Music*, Natural Computing Series, Springer Verlag, Berlin 2008.

computer scientist Simon Colton, which is able to paint urban and rural land-scapes from certain criteria provided by the user.[20] Bulgarian-Canadian artist Krasi Dimtch (stage name Krasimira Dimtchevska) created—together with Canadian programmer Svillen Ranev—software that converts automatically generated sentences into abstract works of art.[21]

If you like dreamlike images, then you must turn to DeepDream, a programme developed by Google in 2015 that can track human faces and transfigure them into psychedelic versions of themselves.[22] More recently, Open AI—a private laboratory in San Francisco dedicated to artificial intelli-gence—created DALL-E and DALL-E 2, two machine learning-based models capable of generating realistic or unrealistic artistic images of all kinds from descriptions based on natural language. The name DALL-E derives from the mixture of Pixar's robotic character WALL-E and the Spanish surrealist painter Salvador Dalí.[23]

In short, it seems that, although we cannot say that machines are truly capable of creating works of genius—original and useful at a very high level—they do seem to be able to do something. Of course, what we have told you represents a drop in the ocean, and there are many projects that aim to attribute creative capabilities to computers. There are even those who have spoken, in this regard, of Artificial Imagination, a term that is capable of encompassing everything we have outlined above—art, music, poetry, humour, and whatnot—in a single system that promises one day to stand alongside its human version.[24]

8.2   Machines that Discover

So, we have ascertained that a machine can produce art. The human imprint is clearly evident, so, for now, we are not dealing with a robotic version of Leonardo or Raphael. Perhaps one day, if computers ever achieve self-awareness, it will be possible to talk about artificial artistic inspiration, robotic creativity, and the ability of machines to appreciate something from an aesthetic point of view. For now, we welcome all these—admittedly impres-sive—developments with some scepticism. At this point, however, let us ask

---

[20] http://www.thepaintingfool.com/.
[21] https://www.krasidimtch.ca/.
[22] https://deepdreamgenerator.com/.
[23] https://openai.com/.
[24] See Aleksander, Igor, *How to Build a Mind: Toward Machines with Imagination*, Columbia University Press, New York 2000.

ourselves: what about science? Can—or will—machines ever make scientific discoveries in total autonomy? After all, scientific discovery is a procedure, right? And the philosophy of science has been working on its formalisation for a long time, so why not think about the possibility of automating it, thus creating real computers and robot scientists, capable of discovering new things and producing paradigm shifts just as the great geniuses of science do? This is exactly what must have occurred to the creators of the first 'discovery programmes', sophisticated software based not on algorithms—i.e., rigid calculation programmes, useful only for solving trivial questions—but on heuristics—i.e., rules of thumb of a certain vagueness, which nonetheless allow creative solutions to non-banal problems to be found, even though they obviously give no guarantees. In practice, heuristics are useful, but the achievement of the result is not guaranteed, whereas with algorithms, the results are certain but predictable. And so, over the years, various research groups have developed a number of such programmes, attributing to them the names of famous scientists: OCCAM, BACON, GALILEO, GLAUBER, HUYGENS, STAHL, FAHRENHEIT, BLACK and DALTON.[25] This is no coincidence: such software is often based on heuristics inspired by the working habits of the scientist from whom it took its name. And so, just as Francis Bacon, in his *Novum Organum* (1620), emphasised the centrality of induction in scientific discovery, BACON relied on the mass of data provided to him to induce—or rather, re-induce—Kepler's third law of planetary motion in a heuristic manner, thus at least demonstrating the reproducibility of a brilliant discovery. Note, however: BACON did not discover the law in question, but only rediscovered it, so we understand that something is still amiss in these attempts to endow computers with scientific genius. In the same way, he rediscovered Joseph-Louis Gay-Lussac's law of combining volumes of gases, Joseph Black's principle of temperature equilibrium, Georg Ohm's laws of current and resistance, Amedeo Avogadro and Stanislao Cannizzaro's method of deriving atomic weights, and more. Things have changed somewhat in recent years. Artificial intelligence has developed further, and Deep Learning has come into play; a more sophisticated version of Machine Learning, it is capable of learning with full or partial supervision, or entirely without supervision. The term 'deep' refers to the fact that it is based on a network structured in several layers, which allows the machine to 'generalise' in a progressive manner, i.e., to extract aspects at an

---

[25] See Bradshaw Gary F., Langley Patrick W., Simon Herbert A., *Studying Scientific Discovery by Computer Simulation*, Science, vol. 222, 1983, pp. 971—977. See also: Shrager, Jeff & Langley, Pat, *Computational Models of Scientific Discovery and Theory Formation*, Morgan Kaufmann, Burlington 1990. Also: Bradshaw Gary F., Langley Patrick W., Simon Herbert A., Zytkow, Jan M., *Scientific Discovery: Computational Explorations of the Creative Process*, The MIT Press, Cambridge 1987.

increasingly abstract level. To give an example: if we use Deep Learning to analyse an image, we will start by extracting more 'basic' aspects such as the contours of the image in question, then, gradually, we will begin to extract more relevant aspects for human observers—symbols, numbers, letters, faces, and so on. The applications of Deep Learning are many. We mention automatic speech recognition; image recognition—aka computer vision—which has proved capable of surpassing humans themselves in this specific task; Neural Style Transfer, i.e., the ability to understand the style of a certain work of art and transfer it, applying it to any photograph or video; language translation; the discovery of new drugs, by simulating and predicting the effects of this or that molecule; medical informatics, e.g., to study the health of a patient from data provided by instruments and wearables; the analysis of medical images for diagnostic purposes; the training of military robots to perform new tasks; and much more. Let us now look at a concrete example, taken from the field of physics: recently, a programme developed by a team at Columbia University in New York and based on Deep Learning has, in fact, discovered a new physics, an alternative to that studied in textbooks. Specifically, Hod Lipson, Boyuan Chen and their colleagues at Columbia's Creative Machines Lab provided their programme with a number of videos of certain physical phenomena occurring here on Earth, and it deduced variables different from those used by established physics. None of this, of course, means that the physics we possess is wrong; on the contrary, it is supported by strong experimental evidence. To make things clearer, let us consider Einstein's famous equation $E = MC^2$. It contains three variables, namely, Energy, Mass, Velocity. Obviously, Enstein did not invent these variables, but inherited them from the tradition of earlier physics. At this point, US scholars wondered whether, starting from the physical phenomena and laws studied by human scientists, it would be possible through their programme to trace back to the same variables, or even to variables other than those already in use, unexpected ones. And that is exactly what happened: the machine, having studied the phenomena, deduced the presence of a number of—unnamed—variables other than the usual, unspecified ones. The programme in question could be used to study phenomena of a cosmological, biological and high-energy physics type, for which the available data are numerous but the theoretical apparatus is still insufficient.[26]

At this point, it is urgent to make a few remarks on all of these rediscovery programmes—our own Simonton made them a few years ago, but, in

---

[26] *Automated discovery of fundamental variables hidden in experimental data* by Boyuan Chen, Kuang Huang, Sunand Raghupathi, Ishaan Chandratreya, Qiang Du and Hod Lipson, 25 July 2022, Nature Computational Science.

our opinion, they also apply to more recent AI developments, such as Deep Learning: to begin with, these programmes do not reproduce the cognitive processes that occurred in the minds of the original discoverers, as far as we know based on the accounts of historians of science; secondly, they do not include all the social and cultural aspects that led to such discoveries; finally, these programmes solved problems, when, in fact, a great scientist is also, and above all, distinguished by their ability to identify the problems in the first place.[27]

8.3   Machines that Transcend

At this point, readers will say: OK, machines do a lot of things, even amazing things; where, however, is their genius? Where is the proof that the synthetic minds under development can reach a level that is not only human, but typical of the most gifted of human beings? At present, we confess, there is no such proof. However, it cannot be ruled out that, one day, robotic genius will become a reality. This is the hope of various theorists of artificial intelligence; that is, the idea that, as research continues, we are destined to come ever closer to the development of a synthetic mind capable not only of achieving human-like self-awareness, but also an intellect superior even to that of the most gifted humans. The Swedish philosopher Nick Bostrom, author of a very fascinating, albeit rather complex, text, *Superintelligence*, is one of the authors of this hypothesis.[28] In it, the author argues that, in the event that the AIs we are already working on now manage to surpass human intellect, they will eventually replace us as the dominant species on planet Earth. Especially in the event that they are endowed with the ability to progressively and spontaneously self-improve themselves. It is not clear, according to Bostrom, when we will ever reach human-like artificial intelligence—whether within a few years, at the end of the century, or later. Once such a level of intelligence is established, however, it is likely that the machines in question will learn to self-perfect, achieving what the philosopher calls 'superintelligence': that is, a cognitive capacity that far exceeds the cognitive performance of humans in virtually all fields of knowledge. Such a super-intelligence would obviously be difficult, if not impossible, to control—who would be able to fool someone who is constitutionally more astute than themselves? The goals of a super-intelligence would vary widely, but it is logical to think that it would

---

[27] Simonton, Dean K., *Genius*, in: Holyoak, Keith J. & Morrison, Robert G. (editors), *The Oxford Handbook of Thinking and Reasoning* pp. 492–509, Oxford University Press, Oxford 2012.
[28] Bostrom, Nick, *Superintelligence: Paths, Dangers, Strategies*, Oxford University Press, Oxford 2014.

end up spontaneously generating what Bostrom calls 'instrumental goals,' such as survival, coherence, cognitive enhancement and acquisition of useful resources. Which means that our artificial super-genius could, firstly, resist any attempt to shut it down, and, secondly, could develop goals incompatible with human presence on the planet. Bostrom's book was a success and, in 2014, it came 17th on the New York Times' list of science bestsellers. Even Elon Musk—who argued that artificial intelligence could be more dangerous than nuclear weapons—and Bill Gates heaped praise on it. Having said this, we would add that the advent of an AI superior to any human genius that has ever existed is still a long way off, and that so-called intelligent machines are only such within certain rather narrow limits, but that this does not mean that it does not make sense to ask the questions that Bostrom poses, and perhaps take appropriate countermeasures.

Concluding Note

Maybe It Does Take a Genius…

To sum up everything that has been discussed so far, we could say that the human genius, as the pinnacle of evolution and the highest expression of the human brain, protects its mystery.

Of course, we have realised that it is not just a romantic concept to be forgotten, that it does not depend on intelligence, that it could have an evolutionary origin—that is, that it could be the fruit of a Darwinian machine for the mass production of ideas—that it is a product of both nature and nurture, that it has some complex connection with madness, and that it cannot—for the moment—be reproduced by machines.

The theme of genius, however, upon deeper analysis, becomes one with the study of human nature itself, with its reasons, which reason does not know. And so, in the end, we realise that the whole of human history can be seen as a succession of geniuses animated by the long breath of expectation, capable of dedicating their lives to capturing a small portion of the true and the beautiful. Small portions, which we can all enjoy, even if we are not geniuses ourselves. Looking forward to the next, unpredictable chapters in this story of human eminence.

© The Editor(s) (if applicable) and The Author(s), under exclusive license to Springer Nature Switzerland AG 2023

R. Manzocco, *Genius*, Springer Praxis Books, https://doi.org/10.1007/978-3-031-27092-5

Bibliography

Aleksander, I. (2000). *How to build a mind: Toward machines with imagination.* Columbia University Press.

Amabile, T. M. (1983). *The social psychology of creativity.* Springer.

Andreasen, N. C., O'Leary, D. S., Cizadlo, T., Arndt, S., Rezai, K., Watkins, G. L., Ponto, L. L., & Hichwa, R. D. (1995). Remembering the past: Two facets of episodic memory explored with positron emission tomography. *American Journal of Psychiatry, 152*(11), 1576–1585.

Andreasen, N. C. (1987). Creativity and mental illness: Prevalence rates in writers and their first-degree relatives. *The American Journal of Psychiatry, 144*(10), 1288–1292.

Andreasen, N. C. (2005). *The creating brain: The neuroscience of genius.* Dana Press.

Andrews, F. M. (1979). *Aichholzer Georg, scientific productivity, the effectiveness of research groups in six countries.* Cambridge University Press.

Arieti, S. (1976). *Creativity: The magic synthesis* (pp. 325–326). Basic Books.

Arlin, P. K. (1990). Wisdom: The art of problem finding. In R. J. Sternberg (Ed.), *Wisdom: Its nature, origins, and development* (Cambridge University Press, New York 1990, pp. 230–243).

Bain, A. (1855). *The senses and the intellect* (Vol. 1977). University Publications of America.

Baltes, P. B., & Smith, J. (1990). *Toward a psychology of wisdom and its ontogenesis.* In R. J. Sternberg (Ed.), *Wisdom: Its nature, origins, and development* (pp. 87–120). Cambridge University Press.

Baltes, P. B., Staudinger, U. M., Maercker, A., & Smith, J. (1995). People nominated as wise: A comparative study of wisdom-related knowledge. *Psychology and Aging, 10,* 155–166.

© The Editor(s) (if applicable) and The Author(s), under exclusive license to Springer Nature Switzerland AG 2023
R. Manzocco, *Genius,* Springer Praxis Books,
https://doi.org/10.1007/978-3-031-27092-5

Baron-Cohen, S. (2020). *The Pattern seekers—How autism drives human invention.* Basic Books.

Barron, F. (1969). *Creative person and creative process, Holt* (p. 42). Rinehart & Winston.

Barron, F. (1963). The needs for order and disorder as motives in creative activity. In C. W. Taylor & F. Barron (Ed.), *Scientific creativity* (pp. 153–160, Wiley, New York 1963).

Bass, B. M., & Riggio, R. E. (2006). *Transformational leadership.* Psychology Press.

Binsted, K., Pain, H., & Ritchie, G. (1997). Children's evaluation of computer-generated punning riddles. *Pragmatics & Cognition, 5*(2), 305–354.

Bloch, F., Hansen, W. W., & Packard, M. (1946). The nuclear induction experiment. *Physical Review, 70*(7), 474–485.

Bloom, B. S. (1963). Report on creativity research by the examiner's office of the university of Chicago. In: C. W. Taylor & F. Barron, *Scientific Creativity: Its recognition and development* (pp. 251–264, Wiley, New York 1963).

Bloom, B. (1985). *Developing talent in young people.* Ballantine Books.

Boden, M. (2004). *The creative mind: Myths and mechanisms.* Routledge, London.

Boring, E. G. (1923). Intelligence as the tests test it. *New Republic, 36*, 35–37.

Bostrom, N. (2014). *Superintelligence: Paths.* Strategies, Oxford University Press, Oxford.

Bouchard, T. J., Jr., & Margaret, H. (1970). Size, performance, and potential in brainstorming groups. *The Journal of Applied Psychology, 54*(1), 51–55.

Chen, B., Huang, K., Raghupathi, S., Chandratreya, I., Du, Q., & Lipson, H. (2022). Automated discovery of fundamental variables hidden in experimental data. *Nature Computational Science.*

Bradshaw, G. F., Langley, P. W., & Simon, H. A. (1983). Studying scientific discovery by computer simulation. *Science, 222*, 971–977.

Langley, P., Simon, H. A., Bradshaw, G. L., & Zytkow, J. M. (1987). *Scientific Discovery: Computational Explorations of the Creative Process*, The MIT Press, Cambridge.

Brannigan, A. (1979). The reification of Mendel. *Social Studies of Science, 9*, 423–454.

Bringsjord, S., & Ferrucci, D. (2000). *Artificial intelligence and literary creativity: Inside the mind of BRUTUS, a storytelling machine..* Lawrence Erlbaum Associates, Hillsdale.

Brodetsky, S. (1942). Newton: Scientist and man. *Nature, 150*, 698–699.

Burke, P., & Polymath, T. (2020). *A cultural history from Leonardo da Vinci to Susan Sontag.* Yale University Press.

Campbell, D. T. (1960). Blind variation and selective retention in creative thought as in other knowledge processes. *Psychological Review, 67*, 380–400.

Cannon, W. (1940). The role of chance in discovery. *The Scientific Monthly, 50*(3), 204–209.

Cattell, R. B. (1966). The personality and motivation of the researcher from measurements of contemporaries and from biography. In. C. W. Taylor & F. Barron (Eds.), *Scientific creativity: Its recognition and development.*

Cole, S., & Cole, J. (1973). *Social stratification in science.* University of Chicago Press.

Conger, J., & Kanungo, R. (1988). *Charismatic leadership in organisations.* SAGE Publications.

Cope, D. (2006). *Computer models of musical creativity.* MIT Press, Cambridge.

Corballis, M. C. (2018). Laterality and creativity : a false trail? In R. E. Jung & O. Vartanian (Eds.), *The Cambridge handbook of the neuroscience of creativity* (pp. 50–57). Cambridge University Press.

Corry, L., Renn, J., & Stachel, J. (1997). Belated decision in the Hilbert-Einstein priority dispute. *Science, 278* (5341): 1270–1273.

Csikszentmihalyi, M., & Rathunde, K. (1990). The psychology of wisdom: An evolutionary interpretation. In: R. J. Sternberg (Ed.), *Wisdom: Its nature, origins, and development* (pp. 25–51). Cambridge University Press.

Csikszentmihalyi, M. (1996). *Creativity: Flow and the psychology of discovery and invention.* HarperCollins.

Currey, M. D., & Rituals, C. (2013). *How artists work.* Knopf.

Cziko, G. A. (1998). From blind to creative: In defence of Donald Campbell's selectionist theory of human creativity. *Journal of Creative Behavior, 32*, 192–208.

Darwin, C., & Darwin, F. (1958). *Autobiography and selected letters.* Courier Corporation.

Dennis, W. (1954). Bibliographies of eminent scientists. *Scientific Monthly, 79*, 180–183.

Dillon, P. C., Graham William, K., Aidells, A. L. (1972). Brainstorming on a "hot" problem: Effects of training and practice on individual and group performance, *Journal of Applied Psychology, 56*, pp. 487–490.

Dryden, J. (1926). *Epistle dedicatory of The Rival Ladies*, in: W. P. Ker (Ed.), *Essays of John Dryden* (Vol. i, pp. 1–9), Oxford: Clarendon Press Oxford 1926 (original essay published 1664).

Dunnette, M. D., Campbell, J., & Jaastad, K. (1963). The effect of group participation on brainstorming effectiveness for 2 industrial samples. *Journal of Applied Psychology, 47*, 30–37.

Dweck, C. S. (2006). *Mindset. The new psychology of success.* Random House, New York.

Ehrenwald, J. (1984). *Anatomy of genius.* Human Sciences Press, New York.

Einstein, A. (1960). *Ideas and opinions.* Crown Publishers Inc.

Elkhorne, J. L. (1967). *Edison—The fabulous drone*, in 73 Vol. XLVI, No. 3 (March 1967), p. 52.

Ericsson, K. A., Krampe, R. T., & Tesch-Römer, C. (1993). The role of deliberate practice in the acquisition of expert performance, *Psychological Review 100*(3), 363–406.

Ericsson, K. (2016). *Anders; Pool, Robert, Peak: Secrets from the new science of expertise*. Houghton Mifflin Harcourt Publishing Company.

Eysenck, H. J. (1993). Creativity and personality: A theoretical perspective. *Psychological Inquiry, 4*, 147–178.

Eysenck, H. J. (1995). *Genius: The natural history of creativity*. Cambridge University Press.

Faraday, M. (1859). *Experimental researches in chemistry and physics,* p. 486.

Fauconnier, G., & Turner, M. (2007). Conceptual integration networks. *Cognitive Science, 22*(2), 133–187.

Findlay, A. (1948). *A hundred years of chemistry*. Duckworth.

Finke Ronald, A., Ward Thomas, B., Smith, S. M. (1992). *Creative cognition: Theory, research, and applications,* MIT Press, Cambridge.

Flaherty, A. W. (2018). Homeostasis and the control of creative drive. In R. E. Jung & O. Vartanian (Eds.), *The Cambridge handbook of the neuroscience of creativity,* (Cambridge University Press, Cambridge 2018, pp. 19–49).

Fox Cabane, O. (2012). *The charisma myth. How anyone can master the art and science of personal magnetism,* Portfolio Hardcover, New York.

Frenkel-Brunswik, E. (1949). Intolerance of ambiguity as an emotional and perceptual personality variable. *Journal of Personality, 18*(1), 108–143.

Fulton, O., & Trow, M. (1974). Research activity in American higher education. *Sociology of Education, 47*, 29–73.

Gagné, F. (2005). From gifts to talents. In R. Sternberg, & J. D. Janet (Eds.), *Conceptions of giftedness* (2nd ed., Cambridge University Press, New York 2005, pp. 98–119).

Galanes, P. (2018). *The mind meld of Bill Gates and Steven Pinker*. New York Times.

Galenson, D. W. (2006). *Old masters and young geniuses*. Princeton University Press, Princeton.

Garcia-Vega, C., & Walsh, V., *Polymathy : The Resurrection of renaissance man and the renaissance brain*. In R. E. Jung & O. Vartanian (Eds.), *The cambridge handbook of the neuroscience of creativity,* (Cambridge University Press, Cambridge 2018, pp. 528–539).

Gardner, H., & Minds, C. (2011). *An anatomy of creativity seen through the lives of Freud, Einstein, Picasso, Stravinsky, Eliot, Graham, and Ghandi*. Basic Books.

Gaston, J. (1973). *Originality and competition in science: A study of the British high energy physics community*. University of Chicago Press.

Gervás, P. (2001). An expert system for the composition of formal Spanish poetry, *Journal of Knowledge-Based Systems*, 181–188.

Goertzel, M. G., Goertzel, V., & Goertzel, T. G. (1978). *Three hundred eminent personalities: A psychosocial analysis of the famous*. Jossey-Bass.

Greenberg, D., et al. (2018). Testing the empathizing systemizing (E-S) theory of sex differences and the Extreme Male Brain (EMB) theory of autism in more than half a million people. *Proceedings of the National Academy of Sciences 115*(48), 12152–12157.

Greene, R. (2012). *Mastery*. Viking Adult.

Hadamard, J. (1945). *The psychology of invention in the mathematical field*. Princeton University Press.

Haken, H. (2004). *Synergetic computers and cognition* (2nd ed.). Springer.

Happé, F., & Vital, P. (2009). What aspects of autism predispose to talent? *Philosophical Transactions of the Royal Society B: Biological Sciences 364*(1522):1369–1375.

Helmholtz, H. V. *An autobiographical sketch*. In *Popular Lectures on Scientific Subjects, Second Series*, (Longmans, New York 1898, pp. 266–291).

Helson, R. (1980). The creative woman mathematician. In L. H. Fox, L. Brody, & D. Tobin (Eds.), *Women and the Mathematical Mystique* (pp. 23–54). Johns Hopkins University Press.

Helson, R. (1971). Women mathematicians and the creative personality. *Journal of Consulting and Clinical Psychology, 36*, 210–220.

Holliday, S. G., & Chandler, M. J. (1986). *Wisdom: Explorations in adult competence*. Karger.

Hollingworth, L. S. (1926). *Gifted children: Their nature and nurture*. Macmillan.

Howell, J. M., & Frost, P. J. (1989). A laboratory study of charismatic leadership. *Organizational Behavior and Human Decision Processes, 43*(2), 243–269.

Hudson, L. (1958). Undergraduate academic record of fellows of the royal society. *Nature, 182*, 1326.

Hull, D. L., Tessner, P. D., & Diamond, A. M. (1978). Planck's principle: Do younger scientists accept new scientific ideas with greater alacrity than older scientists? *Science, 202*.

James, W. (1902). *The varieties of religious experience*. Longmans.

Kipling, R. (1937). *Something of myself*, R. & R. Clark, Edinburgh 1937, p. 162.

Kitchener, K. S. & Brenner, H. G. (1990). Wisdom and reflective judgment: Knowing in the face of uncertainty. In J. Sternberg Robert (Ed.), *Wisdom: Its nature, origins, and development*, Cambridge University Press, New York, pp. 212–229.

Ko, Y., & Kim, J. (2008). Scientific geniuses' psychopathology as a moderator in the relation between creative contribution types and eminence. *Creativity Research Journal, 20*, 251–261.

Koestler, A. (1964). *The act of creation*. Hutchinson.

Kohler, W. (1925). *The mentality of apes,* London: Kegan Paul, Trench, Trubner. U.S. Edition 1925 by Harcourt, Brace & World.

Kroeber, A. (1917). The superorganic. *American Anthropologist, 19*, 163–214.

Kroeber, A. L. (1944). *Configurations of culture growth*. University of California Press.

Kuhn, T. S. (1970). *The structure of scientific revolutions* (2nd ed.), University of Chicago Press, Chicago 1970.

Kuhn, T. S. (1977). *The essential tension: Selected studies in scientific tradition and change*. University of Chicago Press.

Lang, D. (1999). *The New secrets of charisma : How to discover and unleash your hidden powers*. McGraw-Hill.

Lehman, H. C. (1953). *Age and achievement.* Princeton University Press.

Ludwig, A. M. (1992). The creative achievement scale. *Creativity Research Journal, 5*(2), 109–119.

Ludwig, A. M. (1995). *The price of greatness: Resolving the creativity and madness controversy.* Guilford Press.

Ludwig, A. M. (1998). Method and madness in the arts and sciences. *Creativity Research Journal, 11,* 93–101.

Lykken, D. T. (1982). Research with twins: The concept of emergenesis. *Psychophysiology, 19,* 361–373.

Lykken, D. T. et al. (1992). Genetic traits that do not run in families. *Emergenesis. American Psychologist, 47,* 1565–1577.

Lykken, D. T. et al. (1993). Heritability of interests: A twin study, *Journal of Applied Psychology, 78,* 649–661.

Lykken, D. T. (1998). The genetics of genius. In A. Steptoe (Ed.), *Genius and the mind* (pp. 15–37). Oxford University Press, Oxford.

Mach, E. (1896). On the part played by accident in invention and discovery. *The Monist, 6,* 161–175.

Machado, P., & Romero, J. (Eds.). (2008). *The art of artificial evolution: A handbook on evolutionary art and music.* Springer, Berlin.

Mahoney, M. J. (1976). *Scientist as subject: The psychological imperative.* Ballinger.

Maslow, A. (1968). *Toward a psychology of being.* Van Nostrand, New York.

Mateja, D., & Heinzl, A. (2021). Towards machine learning as an enabler of computational creativity. *IEEE Transactions on Artificial Intelligence., 2*(6), 460–475.

McLaren, L. (2000). *Was glenn gould autistic?,* Globe and Mail.

McMahon, D., & Fury, D. (2013). *A history of genius.* Basic Books.

Mednick, S. A. (1962). The associative basis of the creative process. *Psychological Review, 69,* 220–232.

Meehan, J. (1981). *TALE-SPIN,* In: R. C. Shank, C. K. Riesbeck (Eds.), *Inside Computer Understanding: Five Programs plus Miniature, Hillsdale,* Lawrence Erlbaum Associates, Hillsdale.

Merton, R. K. (1961). Singletons and multiples in scientific discovery: A chapter in the sociology of science. *Proceedings of the American Philosophical Society, 105,* 470–486.

Miller, G., & Primates, P. (1997). The evolution of adaptive unpredictability in competition and courtship. In A. Whiten & R. Byrne (Eds.), *Machiavellian Intelligence II* (pp. 312–340). Cambridge University Press.

Minsky, M. (1967). Why programming is a good medium for expressing poorly understood and sloppily formulated ideas. In M. Krampen & P. Seitz (Eds.), *Design and planning 2: Computers in design and communication* (pp. 120–125). Hastings House.

Mitroff, I. I. (1974). Norms and counter-norms in a select group of the apollo moon scientists: A case study of the ambivalence of scientists, *American Sociological Review, 39,* 579–595.

Naifeh, S., & Smith, W. (2011). *Gregory, Van Gogh: The Life.* Random House.

Newell, A., Shaw, J. C., & Simon, H. A. (1962). The processes of creative thinking. In H. E. Gruber, G. Terrell and M. Wertheimer (Eds.), *Contemporary approaches to creative thinking* (University of Colorado, CO, US; This paper was presented at the aforementioned symposium. Atherton Press).

Oakes, L., & Charisma, P. (1997). *The psychology of revolutionary religious personalities*. Syracuse University Press.

Olby, R. C. (1979). Mendel no Mendelian. *History of Science, 17*, 53–72.

Ortega, J. (1957). *The revolt of the masses*, Norton, New York.

Park, R. E. (1928). Human migration and the marginal man. *American Journal of Sociology, 33*, 881–893.

Pelz, D. C., & Andrews, F. M. (1976). *Scientists in organisations: Productive climates for research and development*. Institute for Social Research.

Pereira, F. C. (2006). *Creativity and artificial intelligence: A conceptual blending approach*. Applications of Cognitive Linguistics: Mouton de Gruyter, Amsterdam.

PÉrez, R. P. Ý., & Sharples, M. (2001). MEXICA: A computer model of a cognitive account of creative writing. *Journal of Experimental and Theoretical Artificial Intelligence, 13*, 119–139.

Planck, M. (1949). *Scientific autobiography and other papers*. Philosophical Library.

Platt, W., & Baker, R. A. (1931). The relation of the scientific 'hunch' to research, *Journal of Chemical Education, 8*.

Poincaré, H. (1921). *The foundations of science: Science and hypothesis, the value of science, science and method*. Science Press.

Polgar, L. (2017). *Raise a genius!* , Gordon Tisher 2017, Vancouver, https://slatestar codex.com/Stuff/genius.pdf.

Popescu, T., et al. (2019). The brain-structural correlates of mathematical expertise, *Cortex 114*, 140–150.

Post, F. (1994). Creativity and psychopathology: A study of 291 worldfamous men. *British Journal of Psychiatry, 165*, 22–34.

Potts, J. (2009). *A History of charisma*. Palgrave MacMillan.

Price, D., & Science, L. (1963). *Big science*. Columbia University Press.

Purcell, Ed. M., Torrey, H. C., & Pound, R. V. (1946). Resonance absorption by nuclear magnetic moments in a solid. *Physical Review, 69*, 37–41.

Renzulli, J. S. (1986). The three-ring conception of giftedness: A developmental model for creative productivity. In R. J. Sternberg & J. E. Davidson (Eds.), *Conceptions of giftedness* (pp. 53–92). Cambridge University Press.

Blackburn, R. T., Behymer C. E. & Hall, D. E. (1978). *Correlates of Faculty Publications*, Sociology of Education Vol. 51, No. 2 (Apr., 1978), pp. 132–141.

Roberts, N. C., & Bradley, R. T. (1995). *Limits of Charisma*. In J. A. Sonnenfeld (Ed.) *Concepts of Leadership,* Aldershot, Dartmouth.

Robinson, A. (2010). *Sudden genius?* Oxford University Press, Oxford.

Roe, A. (1952a). A psychologist examines 64 eminent scientists. *Scientific American, 187*(5), 21–25.

Roe, A. (1952b). *The making of a scientist*, Dodd, Mead, New York.

Rogers, C. R. (1954). Toward a theory of creativity. *ETC: A Review of General Semantics, 11*, 249–260.

Rothenberg, A. (1979). *The emerging goddess: The creative process in art, science, and other fields.* University of Chicago Press.

Rushton, J., Murray, H., & Paunonen, S. V. (1983). Personality, research creativity and reaching effectiveness in university professors. *Scientometrics, 5*, 93–116.

Sakakibara, A. (2014). A longitudinal study of the process of acquiring absolute pitch: A practical report of training with the 'chord identification method. *Psychology of Music, 42*(1).

Sanders, J. T. (2000). *Charisma, converts, competitors: Societal and sociological factors in the success of early christianity.* SCM Press.

Schiffer, I. (1973). *Charisma: A psychoanalytical look at mass society.* University of Toronto Press.

Schilpp, P. A. (Ed.), *Albert Einstein: Philosopher-Scientist*, The Library of Living Philosophers, Vol. VII, Evanston, Illinois 1949.

Schlesinger, A. M., Jr. (1964). *The politics of hope.* Eyre & Spottiswoode.

Schou, M. (1979). Artistic productivity and lithium prophylaxis in manic-depressive illness. *British Journal of Psychiatry, 135*, 97–103.

Shrager, J., & Langley, P. (1990). *Computational models of scientific discovery and theory formation.* Morgan Kaufmann.

Herbert, S. (1956). Rational choice and the structure of the environment. *Psychological Review, 63*(2), 129–138.

Simon, H. A., & Chase, W. G. (1973). Skill in Chess: Experiments with chess-playing tasks and computer simulation of skilled performance throw light on some human perceptual and memory processes. *American Scientist, 61*(4), 394–403.

Simon, R. J. (1974). The work habits of eminent scholars. *Sociology of Work and Occupations, 1*(3), 327–335.

Simonton, D. K. (1991). Career landmarks in science: Individual differences and interdisciplinary contrasts. *Developmental Psychology, 27*, 119–130.

Simonton, D. K. (1976a). Do Sorokin's data support his theory? A Study of Generational Fluctuations in Philosophical Beliefs, *Journal for the Scientific Study of Religion, 15*, 187–198.

Simonton, D. K. (1976b). Interdisciplinary and military determinants of scientific productivity: A cross-lagged correlation analysis. *Journal of Vocational Behavior, 9*, 53–62.

Simonton, D. K. (1979). Multiple discovery and invention: Zeitgeist, genius or chance? *Journal of Personality and Social Psychology, 35*, 1603–1616.

Simonton, D. K., & Genius, S. (1988). *A psychology of science.* Cambridge University Press.

Simonton, D. K. (1994). *Greatness.* The Guilford Press, New York.

Simonton, D. K. (1999). *Origins of Genius—Darwinian Perspectives on Creativity.* Oxford University Press.

Simonton, D. K. (2012). *Genius*. In Holyoak, Keith J. & Morrison, Robert G. (editors), *The Oxford Handbook of Thinking and Reasoning* (pp. 492–509, Oxford University Press, Oxford 2012).

Simonton, D. K. (2018). *The genius checklist*. The MIT Press.

Skinner, B. F. (1959). *A case study in scientific method*. In S. Koch (Ed.), *Psychology: A Study of a Science* (Vol. 2), McGraw-Hill, New York 1959, pp. 359–379.

Sorokin, P. (1963). *A long journey; the autobiography of pitirim A*. College and University Press, New Haven.

Staudinger, U. M., Lopez, D. F., & Baltes, P. B. (1997). The psychometric location of wisdom-related performance: Intelligence, personality, and more?. *Personality & Social Psychology Bulletin, 23*, 1200–1214.

Sternberg, R. J., & Lubart, T. I., *Defying the crowd: Cultivating creativity in a culture of conformity*, Free Press; New York 1995.

Sternberg, R. J. (2003). *Wisdom, intelligence, and creativity synthesized*. Cambridge University Press.

Stock, O., & Strapparava, C. (2003). HAHAcronym: Humorous agents for humorous acronyms, *Humor: International Journal of Humour Research, 16*(3), 297–314.

Sulloway, F. J. (1997). *Born to Rebel: Birth order, family dynamics, and creative lives*. Vintage.

Tannenbaum, A. (1983). *Gifted children' psychological and educational perspectives*. Macmillan, New York.

Terman, L. M. (1955). Are scientists different? *Scientific American, 192*, 25–29.

Tinholt, H. W., & Nijholt, A. (2007). Computational humour: Utilizing cross-reference ambiguity for conversational jokes. In *Applications of Fuzzy Sets Theory: 7th International Workshop on Fuzzy Logic and Applications, WILF 2007*, Camogli, Italy, July 7–10, 2007. Proceedings 7 (pp. 477–483). Springer Berlin Heidelberg.

Todd, P. M. (1989). A connectionist approach to algorithmic composition. *Computer Music Journal, 13*(4), 27–43.

Toynbee, A. J. (1946). *A study of history*. Oxford University Press.

Authors, V. (1921). Intelligence and its measurement: A symposium. *Journal of Educational Psychology, 12*(3), 123–147.

Wallas, G. (1926). *The art of thought*. Solis Press.

Walters, J., & Gardner, H. (1984). *The crystallizing experience: Discovering an intellectual gift*. Harvard University.

Wiener, N. (1953). *Ex-Prodigy: My childhood and youth*. Simon & Schuster.

Wiggins, G. (2006). A preliminary framework for description. *Analysis and Comparison of Creative Systems, Journal of Knowledge Based Systems, 19*(7), 449–458.

Willerman, L., Schultz, R., Rutledge, J. N., & Bigler, E. D. (1991). In vivo brain size and intelligence. *Intelligence, 15*(2), 223–228..

Wilson, B. R. (1975). *The noble savages : The primitive origins of charisma and its contemporary survival*. University of California Press Berkeley.

Wordsworth, W. (2010). *The prelude*. Penguin.

Yang, S.-Y., & Sternberg, R. J. (1997). Conceptions of intelligence in ancient Chinese philosophy. *Journal of Theoretical and Philosophical Psychology, 17*(2), 101–119.

Yuasa, M. (1962). The shifting centre of scientific activity in the West. *Japanese Studies in the History of Science, 1*, 57–75.

Zuckerman, H. (1977). *Scientific elite*. Free Press.

Ingram Content Group UK Ltd.
Milton Keynes UK
UKHW010725070423
419815UK00009B/349

9 783031 270918